Routledge Revivals

KEY TO EXERCISES IN LOGIC AND SCIENTIFIC METHOD

KEY TO EXERCISES IN LOGIC & SCIENTIFIC METHOD

BY

A. WOLF

Routledge
Taylor & Francis Group

First published in 1926 by George Allen and Unwin Ltd.

This edition first published in 2018 by Routledge
2 Park Square, Milton Park, Abingdon, Oxon, OX14 4RN
and by Routledge
52 Vanderbilt Avenue, New York, NY 10017, USA

Routledge is an imprint of the Taylor & Francis Group, an informa business

Publisher's Note

The publisher has gone to great lengths to ensure the quality of this reprint but points out that some imperfections in the original copies may be apparent.

Disclaimer

The publisher has made every effort to trace copyright holders and welcomes correspondence from those they have been unable to contact.

A Library of Congress record exists under ISBN:

ISBN 13: 978-0-367-18259-5 (hbk)
ISBN 13: 978-0-367-18260-1 (pbk)
ISBN 13: 978-0-429-06036-6 (ebk)

KEY TO EXERCISES IN LOGIC AND SCIENTIFIC METHOD

GEORGE ALLEN & UNWIN LIMITED
PUBLISHERS LONDON
RUSKIN HOUSE

KEY TO EXERCISES IN LOGIC & SCIENTIFIC METHOD

BY

A. WOLF, M.A., D.LIT.

Professor of Logic and Scientific Method in the University of London;
Fellow of the University of London, University College;
Formerly Fellow of St. John's College, Cambridge;
Author of "Studies in Logic," "Essentials of Scientific Method," etc.

LONDON: GEORGE ALLEN & UNWIN LTD.
RUSKIN HOUSE, 40 MUSEUM STREET, W.C.1

First published . 1926

Printed in Great Britain by
Unwin Brothers, Ltd., Woking

PREFACE

THIS little book was written in the first instance for the benefit of those students of Logic and Scientific Method who receive insufficient or no help in the way of oral instruction. But it may be used profitably even by those who enjoy the advantages of the most efficient teaching.

The main function of the following pages is to deal with the *exercises*, as distinguished from the book-questions. So far as the more elementary book-questions are concerned, the books in ordinary use should supply the required answers. In the case of all the more difficult questions, references are given to various works where the necessary material will be found, and other suggestions are made here and there. But the *exercises* (to the number of over four hundred) are treated, it is hoped, with sufficient fulness.

The range of questions and exercises dealt with in this and its companion volume [1] is so extensive that all sorts and conditions may find some help in them—candidates for Honours Degrees and for the Higher Civil Service, as well as absolute beginners.

[1] *Exercises in Logic and Scientific Method*, New Edition (1926).

The amount of help to be derived from these pages must needs depend on the proper use of the aids they afford. In the study of Logic and Scientific Method, as of everything else, the amount of good to be got out of it depends on the amount of work put into it, and the degree of interest brought to bear upon it. Solutions afford no insight unless the problems are first felt and grappled with.

A. WOLF.

LONDON, *July* 1926.

CONTENTS

GROUP A

SCOPE OF LOGIC

6. J. S. Mill: *System of Logic,* Introduction.
7.* J. Dewey: *How We Think*; or W. B. Pillsbury: *Psychology of Reasoning.*
8.* W. E. Johnson: *Logic,* Vol. I, Introduction.
9.* H. Sidgwick: *Philosophy,* Ch. V.
F. H. Bradley: *Principles of Logic,* Vol. II, (Terminal) Essay I.
10.* B. Bosanquet: *Implication and Linear Inference,* Ch. VII.

* Some of the more difficult subjects are marked with an asterisk.

GROUP B

TERMS

4. J. N. Keynes: *Formal Logic*, Ch. II.
7. H. W. B. Joseph: *Introduction to Logic*, Ch. II.
9. " Abstract " is used here in a wide sense so as to include what is " general " (and concrete) as well as what is abstract in the stricter sense of the term. The wider usage was common formerly, on the ground that general terms are obtained by a process of abstraction.
11. The question whether a term is *used collectively* is quite different from the question whether it *is collective*. Whether a term *is collective* depends on its definition—any name which (in the singular) denotes some group of similar units as a group (not a class or kind) is collective. No context is necessary to determine that. But any name (whether collective or not) can be used in the plural either collectively or distributively, according as what is predicated is applied to the plural subject as one group, or to each separately. How a term is used can only be determined

by reference to the context. In the quotation, the words "all men find their own good" means "each man finds his own good"—"all men" is used distributively; the expression "all men's good" means "the good of mankind as a whole"—"all men" is used here collectively; in the second line the use of "all men" in "all men join in noble brotherhood" may not be so clear, in a sense "any man" may join a brotherhood, but strictly speaking it is only "all men together" (mankind as a whole) that can constitute the brotherhood referred to—so "all men" is here used collectively.

12. H. W. B. Joseph: *Introduction to Logic*, Ch. II.

13. Terms are incompatible when they cannot be predicated of the same subject at the same time.

Contradictory terms are a pair of incompatible terms which between them exhaust all possibilities in some universe of discourse—so that one of them must be true of any relevant subject.

Contrary terms are incompatible terms which express extreme opposites, so that neither of them need be true of any given subject. Terms may be incompatible without being either contrary or contradictory, e.g. "red" and "blue."

14. An expression is ambiguous when it has two

or more meanings or shades of meaning, such that a proposition in which the expression occurs may be true and relevant in one meaning, but not in another.

Some is used in the sense of *some at least*, but also in that of *some only*.

All is used *distributively*, also *collectively*, and sometimes as an adverb for *entirely*.

Or is sometimes used in a sense which makes the alternatives mutually exclusive, and sometimes not.

15.

	Contrary.	*Contradictory.*
(*a*)	The last (or least) servant of the State.	Other than the first servant of the State.
(*b*)	Principles of lowest rank.	Derivative principles.
(*c*)	None.	Elementary studies.
(*d*)	None.	Neither superior nor inferior (equal).
(*e*)	None.	Not equal.
(*f*)	Both less profitable and less honourable.	Either not more profitable or not more honourable.

GROUP C

PREDICABLES, CATEGORIES, ETC.

1. }
2. } H. W. B. Joseph: *Introduction to Logic*, Ch. IV.

3.

	Genus Proximum.	*Differentia.*
Dictionary.	An alphabetically arranged work of reference.	Containing the words of a language (or of some special department of knowledge) and their meaning.
Encyclopedia.	An alphabetically arranged work of reference.	Containing information on all subjects, or on some group of subjects.
Bank.	A financial institution.	For receiving deposits, borrowing and lending money.
Bill of Exchange.	A financial document.	In which one person requests a second to pay money to a third person.
Science.	Systematized knowledge.	Relating to the laws and general characteristics of some class of facts.
Logic.	Science.	Of the general conditions of valid inference.

	Genus Proximum.	*Differentia.*
Economics.	Science.	Of the production, preservation, and distribution of wealth.
Democracy.	System of State government.	Which is controlled by the people.
Socialism.	A social theory.	Which advocates that all land and capital should be owned, and all industries should be controlled, by the people collectively.
Syndicalism.	A social theory.	Which advocates that each industrial group of workers should control the instruments of production which it uses.

4. (*a*) Remote genus [the *genus proximum* would be "lines *in the same plane*"], and *proprium* [the *differentia* would be "equidistant"].
   (*b*) Accident.
   (*c*) Differentia.
   (*d*) Accident.
   (*e*) Proprium.
   (*f*) Proprium.
   (*g*) Accident.

5.–9. H. W. B. Joseph: *Introduction to Logic*, Chs. IV and V

10. Let *n* stand for native, *m* for male, *a* for able-bodied, *s* for skilled, then we get the following sixteen classes :

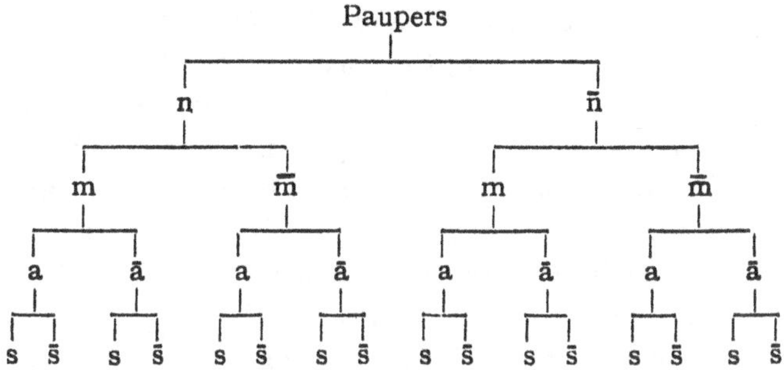

When describing each of the ultimate divisions, the wider classes in which it is included must, of course, be taken into account. Thus the first s stands for the division s a m n (skilled, able-bodied, male, native paupers), while the last s̄ stands for the division s̄ ā m̄ n̄ (unskilled, non-able-bodied, female, foreign paupers).

11. Let *m* stand for many-worded terms, *a* for abstract terms (ā will ∴ = concrete terms), *g* for general term, *c* for collective terms, and *r* for relative terms (r̄ will ∴ = absolute terms), then we get the following scheme :

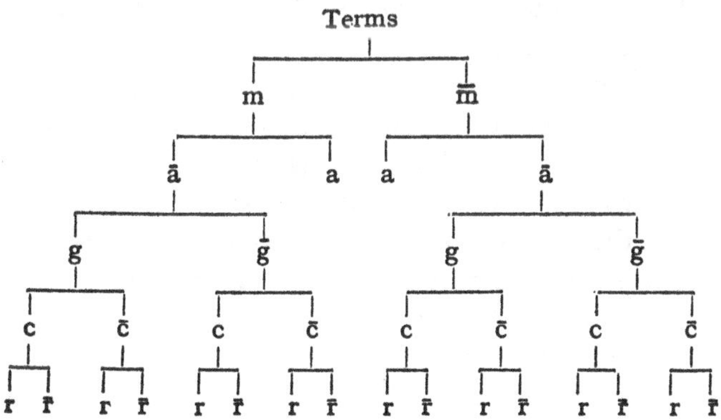

12. The two principles are (*a*) the number of those who rule, and (*b*) the way in which they rule. Two or more principles of division may be employed *successively* (see preceding tables of division, e.g.), but not at the same stage of division.

13. Popularly the term "category" is used for any class, or description, of things; and "predicament" is used for any state or condition, especially for an awkward plight. In Logic, and in Philosophy generally, these terms are used as synonyms, and are applied only to the *ultimate* modes of being, and to our ultimate modes of apprehending reality or of thinking about it. (See H. W. B. Joseph: *Introduction to Logic*, Ch. III.)

14. A. Wolf: *Essentials of Scientific Method*, Ch. III.

15. 16. 17. When things are classified for some definite practical purpose (as distinguished from that of scientific interest) then many things normally distinguished may be classed together for the purpose in hand. See reference under 14.

18. If, when we think of a genus, we think not only of the characters *common* to all its species, but also of the alternative variations which characterize its several species, then the generic term will connote, as well as denote, so much more than any one of its species. But is this what happens always, or even generally?

F. H. Bradley: *Principles of Logic*, Vol. I, Book I, Ch. VI.

19. The specific term implies all that the generic term implies and more—the specific term being the equivalent of the genus and differentia. Therefore, if the specific term can be affirmed of a subject, the generic term can be (in fact is thereby) affirmed of it. The new proposition will actually assert less than the old one. But, just because the specific term implies more than the generic term, the former cannot be affirmed of a subject instead of the latter without going beyond the evidence.

20. What is affirmed of any genus is *ipso facto* affirmed of all its species, for the genus consists of all its species. It is therefore permissible to substitute for it any of its species. The new proposition will assert less than the original one. But to assert of the genus what is only known of one of its species would be to assert of a whole what is only known of a part, and would go beyond the evidence.

21.* H. W. B. Joseph: *Introduction to Logic*, Ch. III.

22.* J. S. Mill: *System of Logic*, Book I, Ch. III, and Joseph, Ch. III.

23.* }
24.* } H. W. B. Joseph: *Introduction to Logic*, Ch. III.

25.* W. E. Johnson: *Logic*, Vol. I, Ch. VII.

GROUP D

LAWS OF THOUGHT

1.* W. E. Johnson: *Logic,* Vol. I, Ch. XIV.
2.* F. H. Bradley: *Principles of Logic,* Vol. I, Book I, Ch. V.
3.* W. E. Johnson: *Logic,* Vol. I, p. 6 note.
4.* W. E. Johnson: *Logic,* Vol. II, Chs. VIII, IX.
5.* W. E. Johnson: *Logic,* Vol. I, Ch. XIV.

GROUP E

THEORY OF JUDGMENT AND IMMEDIATE INFERENCE

4. B. Bosanquet : *Logic,* Vol. I, Ch. I.

6. (*a*) If British subjects are not British born, then they are naturalized.
   All British subjects who are not British born are naturalized.

   (*b*) If lines are not straight, then they are curved.
   All lines which are not straight are curved.

   (*c*) If candidates are to be approved, then they must satisfy certain conditions. Candidates either satisfy certain conditions or they are not approved.

   (*d*) If a book cannot be reviewed favourably, then it should not be reviewed at all.
   A book should either be reviewed favourably or not reviewed at all.

7. (*a*) Equivalent.
   (*b*) Contradictories.
   (*c*) Contraries.
   (*d*) Contradictories.
   (*e*) Compatible, but neither implies the other.

10. Contradictory propositions which have not the

same subjects and the same predicates are propositions in which contradictory predicates are affirmed of the same subject; contrary propositions, under the same circumstances, are propositions in which contrary predicates are affirmed of the same subject.

12. If a proposition of the form S—P implies a proposition of the form P—S, then the latter is called the converse of the former. If a proposition of the form S—RQ implies a proposition of the form Q—ЯS (where R and Я stand for a pair of correlative terms), then the latter proposition may be called the correlative of the former. (Usually it is described as "immediate inference by converse relation.") The converse of S—RQ would be RQ—S. (RQ = P.)

14. SeP and $\bar{P}$eS are inconsistent. For $\bar{P}$eS = Se$\bar{P}$ = SaP, which is the contrary of SeP.

15. SaP implies $\bar{S}$oP (inverse), or $\bar{S}$i$\bar{P}$, subject to the existence of $\bar{P}$, or to the interpretation of the inferred proposition in a problematic sense: There *may be* things, etc.

16. (*a*) "May be" = is possibly. The proposition is problematic. The probability of its truth is estimated by one's view of its coherence, etc.

(*b*) "May" = can, in the sense that frosts sometimes do occur in June.

(*c*) Similar to (*b*), a man *can* do both.

(*d*) "May" = is permitted or allowed.

(*e*) Similar to (*d*). "May not" = is not allowed, or must not.

17. "P implies Q" means that P cannot be true without Q being true.

(i) If P then Q. (ii) Either Q or not P.

18. Let d stand for an adjective or adjectival clause, and C for a concept expressed in a noun or nominal clause, then Immediate Inference by Added Determinants passes from *S is P* to *dS is dP*, while that by Complex Conception passes from *S is P* to *C of S is C of P*. There is no very important difference between them, and both are subject to the same conditions:

(1) d or C must be relevant to S;

(2) they must have precisely the same meaning when associated with S as when associated with P.

Otherwise the result is either nonsense or fallacious.

19.* B. Bosanquet: *Logic*, Vol. I, Ch. I.

20.* The same and F. H. Bradley: *Principles of Logic*, Vol. I, Book I, Ch. VII.

21.* J. Venn: *Empirical Logic*, Ch. X.
B. Bosanquet: *Logic*, Vol. I, Ch. V.

22.* A. Wolf: *Studies in Logic*, Ch. IV.

23.* B. Bosanquet: *Logic*, Vol. I, Ch. V.

24.* F. H. Bradley: *Principles of Logic*, Vol. I, Book I, Ch. I.

25.* B. Russell: *Introduction to Mathematical Philosophy*, Ch. V.

26.* A. Wolf: *Studies in Logic*, Chs. II, III.

27.* B. Bosanquet : *Logic*, Vol. I, Ch. IX.

28.* B. Russell : *Introduction to Mathematical Philosophy*, Ch. XV.

29.* F. H. Bradley : *Principles of Logic*, Vol. II, Book II, Part II, Ch. IV.

30.* B. Bosanquet : *Logic*, Vol. I, Ch. II.

31.*, 32.*, 33.* Some work on Psychology (G. F. Stout's *Manual of Psychology*, e.g.) ; B. Bosanquet : *Logic*, Vol. I, Ch. II ; F. H. Bradley : *Principles of Logic*, Vol. I, Book I, Ch. I. and notes.

34.*, 35.* A. Wolf : *Studies in Logic*, Ch. III.

GROUP F

EXERCISES IN IMMEDIATE INFERENCE

STUDENTS are strongly urged to make themselves familiar with the symbolic representation of propositions. Beginners are prone to cling to the actual wording of the given sentences, and to try to obtain the new form required direct from the words. This is a mistake. Symbolic forms when mastered make the relationships between propositions much clearer and less liable to misapprehension. And the amount of symbolic skill required for ordinary Logic is really small, and readily mastered with a little effort.

There are a few points which should be noted in this connection. In Algebra x, y, etc., may represent negative as well as positive quantities. Similarly in Logic, S, P, Q, R, etc., may stand for negative as well as for positive terms. The important thing to remember is that S and $\bar{S}$ are a pair of contradictory terms; but they are just as contradictory when S stands for the negative term and $\bar{S}$ for the positive term as when vice versa. Similarly with A, B, C, D, etc., which are here used not for terms but for clauses—for the antecedents or consequents of hypothetical propositions, or for the alternatives of disjunctive propositions. In "If A then C," A may = S is not M, and C may = P is not Q, etc. Bearing all this in mind the terms or clauses of any given proposition can always be represented by S and P, or by A and B, etc., and there is no need to begin straightway with negative symbols, which only make it more difficult for the beginner to recognize its relationship to other propositions.

In the following answers, to avoid unnecessary repetitions, the proper categorical form of each proposition is given under (*a*), its contradictory under (*b*), its contrary under (*c*), its obverse under (*d*), the converse under (*e*), the contrapositive under (*f*), and the inverse under (*g*)—that is to say in the order in which they are asked for in the introductory question. Moreover, only a few of the exercises are worked out completely. Once the correct form of the given proposition is known, one has only to compare it carefully with some corresponding proposition already worked out completely to see what the correct eductions, etc., are. Where any particular form is likely to cause trouble, it is given.

1. (*a*) SaP. All who grumble at their tools are bad workmen.
   (*b*) SoP. Some who grumble at their tools are not bad workmen.
   (*c*) SeP. No one who grumbles at his tools is a bad workman.

(*d*) $Se\bar{P}$. No one who grumbles at his tools is a good workman.

(*e*) PiS. Some bad workmen are those who grumble at their tools.

(*f*) $\bar{P}eS$. No good workmen grumble at their tools.

(*g*) $\bar{S}oP$. Some who do not grumble at their tools are not bad workmen.

2\. (*a*) SaP. All who have ideals are blessed.

(*b*) SoP. Some who have ideals are not blessed.

(*c*) SeP. No people who have ideals are blessed.

(*d*) $Se\bar{P}$. No people who have ideals are unblessed.

(*e*) PiS. Some blessed people are those who have ideals.

(*f*) $\bar{P}eS$. No unblessed people are people with ideals.

(*g*) $\bar{S}oP$. Some people without ideals are not blessed.

3\. (*a*) SaP. All home-keeping youths have homely wits.

(*d*) $Se\bar{P}$. No home-keeping youths have other than homely wits.

(*e*) PiS. Some people who have homely wits are home-keeping youths.

(*f*) $\bar{P}eS$. No people having other than homely wits are home-keeping youths.

(*g*) $\bar{S}oP$. Some people other than home-keeping youths are not people with homely wits.

4\. (*a*) SaP. All strong people are strongest when they stand alone.

(*f*) $\bar{P}$eS. No people who are not strongest when they stand alone are strong.

(*g*) $\bar{S}$oP. Some weak people are not strongest when they stand alone.

5. (*a*) SaP. All seers are solitary.

(*f*) $\bar{P}$eS. No non-solitary people are seers.

6. (*a*) SaP. All birds of a feather flock together.

(*f*) $\bar{P}$eS. No birds which do not flock together are birds of a feather.

(*g*) $\bar{S}$oP. Some birds not of a feather do not flock together.

7. (*a*) SaP. A merry heart goes all the way.

(*f*) $\bar{P}$eS. No heart that does not go all the way is a merry heart.

(*g*) $\bar{S}$oP. Some hearts which are not merry do not go all the way.

8. (*a*) SaP. All laughter is fraught with pain.

[The original sentence expresses this rhetorically. If even our sincerest laughter is fraught with pain, other laughter is sure to be so.]

(*f*) $\bar{P}$eS. What is not fraught with pain is not laughter.

9. (*a*) SaP. All human beings are wise after the event.

[The question is merely rhetorical, and a rhetorical question assumes an assertion in an opposite sense—thus, " Who is not P ? " means " All are P," while " Who is P ? " means " No one is P."]

(*f*) $\bar{P}$eS. No one who is not wise after the event is a human being.

(*g*) $\bar{S}$oP. Some other than human beings are not wise after the event.

10. (*a*) SaP. All who reckon without Providence are people who have to reckon twice.

11. (*a*) SaP. All attempts to govern overmuch are mistakes.

(*g*) $\bar{S}$oP. Some attempts not to govern overmuch are not mistakes.

12. (*a*) SaP. All like things seek each other.

13. (*a*) SaP. All cases in which there are too many cooks are cases in which the broth is spoiled.

14. (*a*) SaP. All things thrown at our heads are disgusting.

[Compare 8 above.]

15. (*a*) SaP. All cases of being busy are cases of being half-way to being happy.

16. (*a*) SaP. All rivers are waters which wind somewhere safe to sea.

[A rhetorical expression. If even the weariest river finds its way to sea, then certainly the others will. Compare 8 and 14.]

17. (*a*) SaP. Finding reasons why other folks should be patient is easy.

(*g*) $\bar{S}$oP. Some things other than finding reasons why, etc., are not easy.

18. (*a*) SiP. Some large empires are empires which breed small citizens.

(*b*) SeP. No large empires are empires, etc.

(*c*) None. Particular propositions have no contraries. [Sub-contraries are entirely different from contraries.]

(*d*) So$\bar{P}$. Some large empires are not empires that breed great citizens.

(*e*) PiS. Some empires which breed small citizens are great empires.

(*f*) None.

(*g*) None.

19. (*a*) SaP. All who have not a dram of folly in their mixture are people who have much worse matter in their composition.

(*f*) $\bar{P}$eS. No people who have not something much worse than folly in their composition are people who have not a dram of folly in their mixture.

20. (*a*) SeP. No reparations are things which can undo the evils of warfare.

(*b*) SiP. Some reparations are things, etc.

(*c*) SaP. All reparations are things, etc.

(*d*) Sa$\bar{P}$. All reparations are things which cannot undo the evils of warfare.

(*e*) PeS. No things which can undo, etc., are reparations.

(*f*) $\bar{P}$iS. Some things which cannot undo the evils of warfare are reparations.

(*g*) $\bar{S}$iP. Some things other than reparations are things which can undo the evils of warfare.

21. (*a*) SeP. No truth and physic are well received.

(*b*) SiP. Some truth and physic are well received.

(*c*) SaP. All truth and physic are well received.

(*d*) Sa$\bar{P}$. All truth and physic are ill received.

(*e*) PeS. No thing well received is truth or physic.

(*f*) $\bar{P}$iS. Some things ill received are truth and physic.

(*g*) $\bar{S}$iP Some things other than truth and physic are well received.

22. (*a*) SeP. No person who is not discouraged is really beaten.

(*b*) SiP. Some persons who are not discouraged are really beaten.

(*c*) SaP. All persons who are not discouraged are really beaten.

(*d*) Sa$\bar{P}$. All persons who are not discouraged are persons who are not really beaten.

(*e*) PeS. No persons who are really beaten are persons who are not discouraged.

(*f*) $\bar{P}$iS. Some who are not really beaten are persons who are not discouraged.

(*g*) $\bar{S}$iP. Some who are discouraged are really beaten.

23. (*a*) SeP. No fullest sympathy is felt by those who are lacking in imagination.

(*d*) Sa$\bar{P}$. The fullest sympathy is felt by those who are not lacking in imagination.

(*e*) PeS. No sympathy felt by those who are lacking in imagination is the fullest sympathy.

(*g*) $\bar{S}$iP. Some imperfect sympathy is felt by those who are lacking in imagination.

24. (*a*) SeP. No two people of a trade are people who agree.

(*g*) $\bar{S}iP$. Some people who are not of the same trade are people who agree.

25. (*a*) SeP. No words without thoughts are words that go to heaven.

(*d*) $Sa\bar{P}$. All words without thoughts are words that do not go to heaven.

(*g*) $\bar{S}iP$. Some words with thoughts are words that go to heaven.

26. (*a*) SeP. No existing thing is a thing that has power over the past.

[The original expresses this rhetorically. If even heaven has no such power, then nothing else has.]

(*f*) $\bar{P}iS$. Some things that have no power over the past are existing things.

(*g*) $\bar{S}iP$. Some non-existing things are things that have power over the past.

[This may be a jesting denial, or may refer to the world of imagination.]

27. (*a*) SeP. No person who cannot look at both sides of a question is a properly educated man.

(*f*) $\bar{P}iS$. Some men who are not properly educated are persons who cannot look at both sides of a question.

(*g*) $\bar{S}iP$. Some persons who can look at both sides of a question are properly educated.

28. (*a*) SeP. Honesty is not a thing that has a substitute.

(*b*) SiP.
(*c*) SaP. } The subject term being virtually singular (really abstract) the proposition has a contradictory but no real contrary. It can only be denied in one way: Honesty is a thing that has a substitute. It is different if one discriminate varieties of honesty, such as private, commercial, political!

(*d*) Sa$\bar{P}$. Honesty is a thing that has no substitute.

(*g*) $\bar{S}$iP. Some thing other than honesty is a thing which has a substitute.

29. (*a*) SiP. Some men are masters of their fate.

(*c*) SeP. No men are masters of their fate.

(*d*) So$\bar{P}$. Some men are not beings who are not masters of their fate (or, some men are not mastered by their fate).

(*b*), (*e*), (*f*), (*g*). None.

30. (*a*) SiP. Some attempts to enforce justice are more troublesome and costly than to bear an injustice.

(*d*) So$\bar{P}$. Some attempts, etc., are not things which are not more troublesome, etc.

(*e*) PiS. Some things more troublesome and costly than bearing an injustice are attempts to enforce justice.

[Not, "Sometimes to bear an injustice is less troublesome and costly than to enforce justice"—this is the correlative. See Group E, No. 12.]

31. (*a*) SiP. Some unconscious cruelty is a result of a lack of imagination.

(*d*) So$\bar{P}$. Some unconscious cruelty is not other than the result of a lack of imagination.

32. (*a*) SiP. Some victories obtained in a wrong cause are the most awful calamity, etc.

33. (*a*) SoP. Some races are not won by the swift.

(*d*) Si$\bar{P}$. Some races are things not won by the swift.

(*f*) $\bar{P}$iS. Some things not won by the swift are races.

34. (*a*) SiP. Some who win most of the battles are those who lose the war.

(*d*) So$\bar{P}$. Some who win most of the battles are not those who win the war.

35. (*a*) SiP. Some defeats on the battlefield are blessings.

(*d*) So$\bar{P}$. Some defeats on the battlefield are not other than blessings.

36. (*a*) SiP. Some hardships are blessings in disguise.

(*d*) So$\bar{P}$. Some hardships are not other than blessings in disguise.

37. (*a*) Si$\bar{P}$. Some roses are born to blush unseen.

(*d*) SoP. Some roses are not born not to blush unseen.

38. (*a*) SiP. Some economic phenomena which are not seen are better worth studying than those which are seen.

(*d*) So$\bar{P}$. Some unseen economic phenomena are not phenomena not better worth studying than those which are seen.

(*e*) PiS. Some economic phenomena better worth studying than those seen are those which are not seen.

[The converse must not be confused with the correlative, which would be in this case : " Some economic phenomena which are seen are less worth studying than those which are not seen."]

39. (*a*) SoP. Some (or most) honest work is not unrewarded.

(*d*) Si$\bar{P}$. Some honest work is rewarded.

40. (*a*) SoP. Some men of violent minds are not lacking in tenderness.

[The expression " are not all lacking " is rather ambiguous. It may mean " some are not (= not all are) lacking " or " are not entirely (all = altogether) lacking."]

(*d*) Si$\bar{P}$. Some men of violent minds are tender-hearted.

41. (*a*) SoP. Some obscure books are not profound.

(*d*) Si$\bar{P}$. Some obscure books are books which are not profound. [Avoid " superficial " and any other term which is not the true contradictory of " profound."]

(*f*) $\bar{P}$iS. Some books which are not profound are obscure.

42. (*a*) SoP. Some (most) people are not able to describe adequately what they see.

(*d*) Si$\bar{P}$ Some people are unable to, etc.

(*f*) $\bar{P}$iS. Some who are unable to describe, etc., are (ordinary) people.

43. (*a*) SoP. Some gases are not liquefiable.

(*d*) Si$\bar{P}$. Some gases are non-liquefiable.

(*e*) $\bar{P}$iS. Some non-liquefiable substances are gases.

44. (*a*) SaP. All poor people are generous in their intentions.

(*d*) Se$\bar{P}$. No poor people are people who are not generous, etc.

(*g*) $\bar{S}$oP. Some people who are not poor are not generous, etc. [Avoid " rich " for " not poor."]

45. (*a*) SiP. Some cheap labour is efficient.

[" Even " insinuates that most cheap labour is, or is held to be, inefficient ; but this is not the main point of the statement.]

46. (*a*) SoP. Some (most) people are not just to those whom they dislike.

(*d*) Si$\bar{P}$. Some people are unjust to those, etc.

(*f*) $\bar{P}$iS. Some who are unjust to those whom they dislike are (ordinary or actual) people.

47. (*a*) SaP. All things out of sight are out of mind.

(*d*) Se$\bar{P}$. No things out of sight are retained in the mind.

(*g*) $\bar{S}$oP. Some things in sight are not out of mind.

48. (*a*) SoP. Some people who are not discontented are not unambitious.

(*d*) Si$\bar{P}$. Some people, etc., are ambitious.

(*f*) $\bar{P}$iS. Some ambitious people are people who are not discontented.

49. (*a*) SiP. Some jokes are worth ten arguments.

(*d*) So$\bar{P}$. Some jokes are not things which are not worth ten arguments.

["Sometimes" is rather ambiguous here. It may mean "some"—the adverb of time being commonly used as a mark of quantity—or it may mean "some occasions." In the latter case the sentence may be better expressed thus: Some occasions are such that a joke is worth ten arguments.]

50. (*a*) SiP. Some attempts to manage other people with the aid of fun are more successful than bullying.

(*d*) So$\bar{P}$. Some attempts, etc., are not attempts which are not more successful than bullying.

[Avoid "are not less successful" which is not the correct contradictory of "are more successful."]

(*e*) PiS. Some attempts to manage other people which are more successful than bullying them are attempts to manage them with the aid of fun.

[Avoid confusion with the correlative, which would be: Bullying is less successful in the management of other people than fun.]

51. (*a*) SaP. The measure of a man's value is his ideals.

(*b*) SoP. The measure of a man's value is not his ideals.

(*c*) SeP. None. The proposition being essentially singular it has no contrary different from its contradictory, unless we introduce contrary *terms* in the predicate.

(*d*) Se$\bar{P}$. The measure, etc., is not other than his ideals.

(*e*) PiS. A man's ideals are the measure of his value.

(*f*) $\bar{P}$eS. Nothing other than his ideals is the measure of a man's value.

(*g*) $\bar{S}$oP. Some things that are not the measure of a man's value are not his ideals.

52. (*a*) SaP. All who venture nothing are people who have nothing.

(*f*) $\bar{P}$eS. No people who have anything are people who venture nothing.

(*g*) $\bar{S}$oP. Some people who venture something are not people who have nothing.

53. (*a*) SaP. All things evil are things that have some soul of goodness.

[More correctly this is a rhetorical way of saying: "There is some soul of goodness in *all* things—*even* in things evil." The proposition should therefore be: "Some goodness is found in all things," or "All things are good in some respect."]

54. (*a*) SoP. Some (most) people are not conscious of their shortcomings.

(*d*) Si$\bar{P}$. Some people are unconscious of their shortcomings.

(*f*) $\bar{P}$iS. Some beings who are unconscious of their shortcomings are human beings (or people).

55. (*a*) SeP. No one who really loved is a person who did not love at first sight.

[A rhetorical question expecting a negative answer.]

(*d*) Sa$\bar{P}$. All who really loved are people who loved at first sight.

(*e*) PeS. No one who did not love at first sight is a person who really loved.

(*g*) $\bar{S}$iP. Some who did not really love are people who did not love at first sight.

56. (*a*) SiP. Some people who have excellent morals are people having odious ways.

(*d*) So$\bar{P}$. Some people who have excellent morals are not people without odious ways.

57. (*a*) SoP. Some who have suffered from oppression are not people who have learned the lesson of not oppressing others.

(*d*) Si$\bar{P}$. Some who, etc., are people who have not learned the lesson, etc.

(*f*) $\bar{P}$iS. Some who have not learned the lesson of not oppressing others are people who have suffered from oppression.

58. (*a*) SaP. Consolation obtained by shutting the eyes to facts is poor.

(*d*) Se$\bar{P}$. Consolation obtained, etc., is not other than poor.

(*f*) $\bar{P}$eS. No other than poor consolation is that obtained by shutting the eyes to facts.

(*g*) $\bar{S}$oP. Some consolation not obtained by shutting the eyes to facts is not poor.

59. (*a*) SiP. Some heroic actions are easier than right actions.

(*d*) So$\bar{P}$. Some heroic actions are not as difficult as right actions.

(*e*) PiS. Some actions which are easier than right actions are heroic actions.

[The correlative would be: "Some right actions are more difficult than heroic actions."]

60. (*a*) SeP. No things which we do not love are things which we can understand thoroughly.

(*d*) Sa$\bar{P}$. All things which we do not love are things which we cannot understand thoroughly.

(*g*) $\bar{S}$iP. Some things which we love are things which we can understand thoroughly.

61. (*a*) SeP. No bad news or good advice is well received.

(*d*) Sa$\bar{P}$. All bad news and good advice are ill-received.

(*e*) PeS. Nothing that is well received is bad news or good advice.

(*g*) $\bar{S}$iP. Something that is not bad news or good advice is well received.

[In some cases "or" is substituted for the original "and" to avoid the suggestion that it must be *both* bad news and good advice.]

62. (*a*) SaP. All the good intentions in the world are things that cannot undo the evil caused by ignorance.

(*d*) Se$\bar{P}$. No good intentions are things that can undo the evil caused by ignorance.

(*g*) $\bar{S}$oP. Some things other than good intentions are not things that cannot undo the evil caused by ignorance.

63. (*a*) SaP. All quarrels are things which need two people.

(*d*) Se$\bar{P}$. No quarrels are things which do not need two people.

(*e*) PiS. Some things which require two people are quarrels.

(*g*) $\bar{S}$oP. Some things other than quarrels are not things which need two people.

64. (*a*) SaP. All who waste not are people who want not.

(*d*) Se$\bar{P}$. No one who wastes not is a person who wants (or is in want).

(*g*) $\bar{S}$oP. Some who waste are not people who do not want.

65. (*a*) SeP. No faint heart is a heart that won a fair lady.

(*d*) Sa$\bar{P}$. All faint hearts are hearts that have not won a fair lady.

(*g*) $\bar{S}$iP. Some hearts that were not faint are hearts that won fair lady.

66. (*a*) SaP. All the perfumes of Arabia are things that will not sweeten this little hand.

(*d*) $Se\bar{P}$. No perfumes of Arabia are things that will sweeten this little hand.

(*g*) $\bar{S}oP$. Some things other than the perfumes of Arabia are not things that will not sweeten this little hand.

67\. (*a*) SaP. All cases of "no news" (or, of absence of news) are cases of good news.

(*d*) $Se\bar{P}$. No cases of "no news" are cases of other than good news.

(*g*) $\bar{S}oP$. Some news is not good news.

68\. (*a*) SaP. All places where there is smoke are places where there is fire.

(*d*) $Se\bar{P}$. No places where there is smoke are places where there is no fire.

(*g*) $\bar{S}oP$. Some places where there is no smoke are not places where there is fire.

69\. (*a*) SeP. No wise men are men who bewail their woes.

(*d*) $Sa\bar{P}$. All wise men are men who do not bewail their woes.

(*g*) $\bar{S}iP$. Some unwise men are men who bewail their woes.

70\. (*a*) SaP. Every stitch in time is a stitch that saves nine.

(*d*) $Se\bar{P}$. No stitch in time is a stitch that does not save nine.

(*e*) PiS. Some (single) stitches that save nine stitches are stitches in time.

(*g*) $\bar{S}oP$. Some stitches not made in time are not stitches that save nine.

71\. (*a*) SaP. To form one's own judgment is more difficult than to receive it from others.

(*d*) Se$\bar{P}$. To form one's own judgment is not so easy as to receive it from others.

(*e*) PiS. Some thing more difficult than to receive the views of others is to form one's own judgment.

[The correlative would be: "To receive it from others is easier than to form one's own judgment."]

(*g*) $\bar{S}$oP. Some things other than forming one's own judgment are not more difficult than to receive it from others.

72. (*a*) SoP. Some normal actions are not right.

(*d*) Si$\bar{P}$. Some normal actions are wrong.

73. (*a*) SoP. Some (most) who have seen the devastation of war are not people who believe in its glory.

(*d*) Si$\bar{P}$. Some who have seen, etc., are people who do not believe in the glory of war.

74. (*a*) SaP. All cases in which ignorance is bliss are cases in which it is folly to be wise.

(*d*) Se$\bar{P}$. No cases in which ignorance is bliss are cases in which it is not folly to be wise.

(*g*) $\bar{S}$oP. Some cases in which ignorance is not bliss are not cases in which it is folly to be wise.

75. (*a*) SoP. Some (most) people who only think of their success are not people who succeed.

(*d*) Si$\bar{P}$. Some people who, etc., are people who do not succeed [or "people who fail"].

76. (*a*) SoP. Some of those who laugh with you when you laugh are not people who weep with you when you weep.

(*d*) $Si\bar{P}$. Some of those who laugh, etc., are people who do not weep with you, etc.

77. (*a*) SaP. Every person whose life is in the right is a person who can't be wrong.

(*d*) $Se\bar{P}$. No person whose life, etc., is a person who can be wrong.

(*e*) PiS. Some people who can't be wrong are people whose life is in the right.

(*g*) $\bar{S}oP$. Some people whose life is not in the right are not people who can't be wrong.

78. (*a*) SoP. Some clever men are not good.

79. (*a*) SoP. Some who desire the good are not willing to pay the price.

(*d*) $Si\bar{P}$. Some who desire the good are unwilling to pay the price.

80. (*a*) SaP. All things noble are as difficult as they are rare.

(*d*) $Se\bar{P}$. No things noble are things which are not as difficult as they are rare.

81. (*a*) All people who fail blame others for it . . . . SaP

Some who fail do not blame others for it . . . . SoP.

Contradictory propositions.

(*b*) Some (most) of those who had worked are not people who failed . . . SoP.

Some who failed are not people who had not worked . . . . . $Po\bar{S}$.

Po$\bar{S}$ = PiS = SiP, which is consistent with SoP, though neither implies the other. Sub-contraries.

(*c*) All experts are people of wide experience . . . . SaP.
No inexperienced people are experts $\bar{P}$eS.
Contrapositives.

(*d*) All diamonds are white . . . . . SaP.
Some diamonds are not white . . SoP.
Contradictories. Or " white " and " yellow " can be treated as incompatible terms affirmed of the same subject, and so the propositions are incompatible—for other than formal reasons of their quality and quantity.

82. (*a*) All workers succeed . . . . . . SaP.
Some workers do not succeed . . SoP
Contradictories. Can also be treated as material contradictories—contradictory predicates being affirmed of the same subject.

(*b*) SaP and PaS. They are consistent, but neither implies the other.

(*c*) All the candidates are graduates . SaP.
None of the candidates is a non-graduate . . . . Se$\bar{P}$.
Obverse.

(*d*) All popular people are considerate SaP.
All who are not considerate are not popular . . . . $\bar{P}$a$\bar{S}$.
Obverted contrapositive.

83. (*a*) No imprudent person is happy . SeP.
(*b*) No prudent people are unhappy . $\bar{S}e\bar{P}$.
(*c*) Some (most) unhappy people are imprudent . . . . $\bar{P}iS$.
(*d*) Some (most) prudent people are not unhappy . . . . $\bar{S}o\bar{P}$.

(*a*) is compatible with (*b*) but does not imply it (SeP only implies $\bar{S}o\bar{P}$); (*a*) implies (*c*)—$\bar{P}iS = Si\bar{P} = SoP$, the subalternate of SeP; (*a*) implies (*d*), its obverted inverse.

(*b*) implies (*c*)—$\bar{S}e\bar{P} = \bar{P}e\bar{S} = \bar{P}aS$; and (*d*) is the subalternate of (*b*)

(*c*) and (*d*) are compatible, but do not imply each other (just like SoP and PoS, for (*c*) $\bar{P}iS = \bar{P}o\bar{S}$, and (*d*) is $\bar{S}o\bar{P}$).

84. The falsity of SeP implies the truth of its contradictory SiP—some treaties can prevent war (i.e. war is preventable).

The truth of SaP (All treaties fail to prevent war) does *not* imply $\bar{S}aP$ (All things other than treaties also fail to prevent war—which would mean that war is unpreventable).

85 (*a*) $SaP = Se\bar{P} = \bar{P}eS = \bar{P}a\bar{S}$.
(*b*) $Pe\bar{S} = PaS$.
(*c*) $Se\bar{P} = \bar{P}eS = \bar{P}a\bar{S}$.
(*d*) $\bar{P}a\bar{S}$.
(*e*) $\bar{S}i\bar{P} = \bar{P}i\bar{S}$.
(*f*) $PeS = SeP$.
(*g*) $\bar{P}i\bar{S} = \bar{S}i\bar{P} = \bar{S}oP$.

(*h*) $PaS = Pe\bar{S}$.
(*i*) $\bar{P}e\bar{S} = \bar{S}e\bar{P} = \bar{S}aP$.
(*j*) $\bar{S}a\bar{P}$.
(*k*) $\bar{P}iS = Si\bar{P} = SoP$.
(*l*) $PiS = SiP$.
(*m*) SiP.
(*n*) $\bar{S}iP = Pi\bar{S} = PoS$.
(i) (*c*), (*d*), (*e*), (*g*), (*l*), (*m*).
(ii) (*c*), (*d*).
(iii) (*f*), (*i*), (*k*).
(iv) (*b*), (*h*), (*j*), (*n*).

86. (*a*) $SiP = PiS$ and $So\bar{P}$.
(*b*) $\bar{P}iS = Si\bar{P} = SoP$.
(*c*) $Si\bar{P}$.
(*d*) $\bar{P}e\bar{S}$.
(*e*) $Sa\bar{P} = SeP$.
(*f*) $\bar{S}aP$.
(*g*) $Pa\bar{S} = PeS = SeP$.
(*h*) $Pe\bar{S} = PaS$.
(*i*) $\bar{S}e\bar{P}$.
(*j*) $PeS = SeP$.
(*k*) $\bar{P}aS$.
(i) None.
(ii) (*h*).
(iii) (*e*), (*g*), (*j*).
(iv) (*b*), (*c*), (*d*), (*f*), (*h*), (*i*), (*k*).

87 (*a*) $SeP = Sa\bar{P}$.
(*b*) $\bar{P}e\bar{S}$.
(*c*) $Pa\bar{S} = PeS = SeP$.
(*d*) $\bar{S}i\bar{P}$.
(*e*) $\bar{S}iP = Pi\bar{S}$.
(*f*) $Sa\bar{P} = SeP$.

(*g*) $\bar{P}$aS.

(i) (*c*), (*e*), (*f*).

(ii) (*c*), (*f*).

(iii) None.

(iv) (*b*), (*d*), (*g*).

88. (*a*) SeP = Sa$\bar{P}$.

(*b*) Sa$\bar{P}$ = SeP.

(*c*) $\bar{S}$i$\bar{P}$ = $\bar{P}$i$\bar{S}$ = $\bar{P}$oS.

(*d*) $\bar{P}$i$\bar{S}$ = $\bar{P}$oS.

(*e*) $\bar{P}$oS = $\bar{P}$i$\bar{S}$.

(*f*) $\bar{P}$aS.

(*g*) $\bar{P}$eS = $\bar{P}$a$\bar{S}$ and Se$\bar{P}$ = SaP.

(*h*) PiS = SiP.

(i) (*b*).

(ii) (*b*).

(iii) (*g*), (*h*).

(iv) (*c*), (*d*), (*e*), (*f*).

89. This is immediate inference by complex conception. S is P, therefore C of S is C of P. Freedom of speech is the essence of democracy, therefore, the renunciation of the one is the renunciation of the other.

90. Let R = rich, and G = good, then there are four possible combinations: G R, G $\bar{R}$, $\bar{G}$ R, $\bar{G}$ $\bar{R}$. Now *ceteris paribus* X would rather be good than bad, that is, he would rather be G R than $\bar{G}$ R, and G $\bar{R}$ rather than $\bar{G}$ $\bar{R}$; but he would rather be rich than good, that is, he would rather be $\bar{G}$ R than G $\bar{R}$. His order of preference is therefore this—G R, $\bar{G}$ R, G $\bar{R}$, $\bar{G}$ $\bar{R}$.

GROUP G

THEORY OF INFERENCE

2.* B. Bosanquet : *Logic,* Vol. II, Ch. I.
L. T. Hobhouse : *Theory of Knowledge,* Part II, Chs. I–VI.

3.* B. Bosanquet : *Implication and Linear Inference,* Ch. I.

4.* F. H. Bradley : *Principles of Logic,* Vol. I, Book II, Part I ; and Vol. II, (Terminal) Essay I.

5.* B. Bosanquet : *Logic,* Vol. II, Ch. I.
F. H. Bradley : *Principles of Logic,* Vol. II, (Terminal) Essay I.

6.* F. H. Bradley : *Principles of Logic,* Vol. II, Book III, Part I, Ch. VI.

7.* B. Bosanquet : *Logic,* Vol. II, Ch. I.

8.* F. H. Bradley : *Principles of Logic,* Vol. II, Book III, Part I, Ch. II.

9.* F. H. Bradley : *Principles of Logic,* Vol. II, (Terminal) Essay II.

GROUP H

SYLLOGISTIC DOCTRINE

3. The conclusion, if any, will have to be negative and distribute its predicate. The major term must therefore be distributed in its premise. But as the minor premise is negative, the major premise must be affirmative, and the only possibility of its distributing the major term is by being universal and having the major term for its subject (PaM).
4. If the conclusion is universal affirmative, both premises must be universal affirmative, and distribute only two terms between them. One of these must be the minor term (otherwise the conclusion must not be universal). Therefore the middle term can only be distributed once. If the conclusion is universal negative, one premise must be universal affirmative and the other universal negative. They will therefore distribute three terms between them. But two of these must be the minor term and the major term (otherwise the conclusion could not be universal and negative). Therefore,

in this case also, the middle term can only be distributed once.

5. (*a*) Most M's are P ; most M's are S ; ∴ SiP.

   (*b*) MeP or PeM.

   Se$\overline{M}$ (= SaM) or So$\overline{M}$ (= SiM)

   ∴ SeP or SoP.

6. In Fig. I only the major premise can be negative (to avoid illicit process of the major term), and then it must be universal (to distribute the middle term). Therefore, an O premise is impossible in Fig. I.

   In Fig. II one premise must be negative (to distribute the middle term), but the major premise must be universal (to avoid illicit process of the major term). Therefore, only the minor premise can be O in Fig. II.

   In Fig. III only the major premise can be negative (to avoid illicit process of the major term). An O premise in Fig. III can therefore only occur as major premise.

   In Fig. IV if either premise is negative the major premise must be universal (to avoid illicit process of the major term). The major premise cannot therefore be O. If the minor premise is O, it will not distribute the middle term. Nor, in that case, can the middle term be distributed in the major premise (which will have to be affirmative). Therefore, in Fig. IV, neither premise can be O.

To sum up, O cannot be a premise in Figs. I

and IV, can only be the minor premise in Fig. II, and the major in Fig. III.

7. (*a*) IEO.

(*b*) AAA in Figs. II–IV ; IAI in Figs. I, II ; OAO in Figs. I, II, IV ; EAE in Figs. III, IV ;
AOO in Figs. I, III, IV.

8. (*a*) When the hypothetical proposition is reciprocal.

(*b*) When the alternatives are mutually exclusive.

9. Let the original dilemma be :

If A, then B ; and if C then D.
But either A or C.
∴ Either B or D.

The rebuttal will then take the form :

If A then not D ; and if C then not B.
But either A or C.
∴ Either not D or not B.

The rebuttal is worthless, as its conclusion is perfectly compatible with the conclusion of the original dilemma, which only demands B *or* D, not B *and* D. The best way of meeting an invalid dilemma is to indicate what alternative possibilities have been omitted.

12. No, because the proof does not rest on any of the properties peculiar to the diagram (such as its actual size, colour, etc.), but on those characteristics which are common to the whole class which it represents.

14. L. T. Hobhouse : *Theory of Knowledge,* Part I, Ch. VI.

16. As only A and E propositions are contraries, the two syllogisms must have an A and an E conclusion respectively. Now the syllogism with the A conclusion can only have the form : MaP ; SaM ; ∴ SaP. The syllogism with the E conclusion may be of the form : MeP ; SaM ; ∴ SeP, or PeM ; SaM ; ∴ SeP. In either case the two syllogisms would have an identical premise, viz. SaM, and contrary conclusions, viz. SaP and SeP. The Figs. and Moods are obviously Fig. I, AAA, and either Fig. I, or Fig. II, EAE.

17.* The premises must all be universal, else they would have no contraries ; and, unless each contained a negative premise, the other would have to contain two negative premises. The following two syllogisms satisfy the required conditions.

PeM ; SaM ; ∴ SeP.
PaM ; SeM ; ∴ SeP.

The possibility of drawing the same inference from pairs of contrary premises casts no suspicion whatever on the validity of syllogistic inference. It is just what should be expected. In a syllogism a negative conclusion is drawn on the ground that the minor term and the major term are related in opposite ways to the same middle term, and must consequently be different from each other. So long as S and P are related in opposite ways to the same M, it makes no difference to the result whether P is related negatively to M, while S is related affirmatively to it, or vice versa.

18.* In every valid syllogism the conclusion is implied in its premises. This means that if the premises are true, then the conclusion must be true; if the conclusion is not true, then the premises cannot both be true. For a syllogism to prove one of its own premises would consequently involve a vicious circle—premise and conclusion would be made to justify each other. For a syllogism to disprove its own conclusion would involve self-contradiction—for its conclusion is what the syllogism proves, and cannot, therefore, also disprove. The syllogisms given as alleged examples of these feats do not really exemplify them at all, and are at best only specious counterfeits of them.

(i) This syllogism presumably is intended to show that a syllogism may prove one of its own premises. The form of the syllogism is: MiP; MaS; ∴ SiP.

It is obvious that the conclusion is *not* identical with either of its premises. Nor is the resemblance between the conclusion and the major premise more than superficial. The conclusion might be true even if the major premise were not. The major premise cannot, therefore, be said to be *proved* by the conclusion in any sense. While to suggest that the *particular* conclusion *proves* the *universal* minor premise would be even more absurd.

(ii) This syllogism is presumably intended to illustrate how a syllogism may disprove its own conclusion. Does it ? The idea which underlies this ingenious conundrum appears to be this. Unless it is allowed that a syllogism really warrants its conclusion, even this conclusion would not follow from these premises. To draw a conclusion at all (even the conclusion that syllogism is not genuine inference) implies that syllogism is regarded as genuine inference, and this assumption contradicts the conclusion actually drawn. But there is nothing in the actual syllogism itself which is inconsistent in any way with (much less *disproves*) the conclusion. Nor is there any essential inconsistency, even implicit consistency, between the conclusion and the use of the syllogistic form. " Genuine " is, and can be, used in the conclusion only in the same sense in which it is used in the major premise, and that sense is different from " valid." The assumption of the " validity " of syllogism (implied in using it at all) is not really inconsistent with the conclusion that syllogism is not inference from the known to the unknown (or "genuine" in this special sense). It is the ambiguous use of "genuine" that makes the example as specious as it is. Express the syllogism symbolically, and the pretence vanishes. PaM ; SeM ; ∴ SeP.

19.* B. Bosanquet: *Implication and Linear Inference*, Ch. VI.

F. H. Bradley: *Principles of Logic*, Vol. II, (Terminal) Essay I.

20.* Facts or actual occurrences are usually the resultants of many forces, some of which resist, mask, or modify others. Science makes its problems manageable by abstraction and simplification. This is legitimate, so long as it is not forgotten that abstraction has been made. But when inferences from partial data, obtained by abstraction and simplification, are applied straightway to concrete facts, without taking into account other complicating tendencies, then disappointment is apt to result. This does not mean that the premises (or "theories" as they are usually called in this connection) were necessarily wrong, only that the procedure was too abstract, that those premises were not the only relevant truths which should have been considered.

GROUP I

SYLLOGISTIC EXERCISES

THE student's attention may be directed again to the note about symbols at the beginning of Group F (page 25).

In the following answers the major premise of a syllogism is generally put first, and the minor premise next; but in the case of chains of syllogisms this is done only in the case of the first syllogism of the chain, in order to avoid the need of repeating statements—the symbolic expressions which accompany the verbal statements should make the major and the minor premises in each syllogism easily recognizable.

1. What is believed universally is true . MaP;
   That all men have an ample measure of common sense is believed universally . . . . SaM;
   ∴ That all men have an ample measure of common sense is true . SaP.
2. All persons should undergo military training . . . . PaM;
   No women need undergo military training . . . . SeM;
   ∴ No women are persons . . . . SeP.

[It may be that an injustice is done to Mr. X by neglecting the ***universe of discourse.*** At a recruiting meeting one might readily use the word "person" in the sense of "male person."]

3. What is true of virtues [in respect of volition] is true of vices . . . . MaP ;
That they are voluntary is true of virtues . . . . SaM ;
∴ That they are voluntary is true of vices . . . . SaP.

4. Some malcontents are selfish . . . MiP ;
All malcontents are radicals . . . MaS ;
∴ Some radicals are selfish . . . . . SiP.
By obversion : Some radicals are not unselfish . . . . So$\bar{P}$.

5. Marcus Aurelius is a good man . . MaP ;
Marcus Aurelius is an emperor . . . . MaS ;
∴ Some emperor is a good man . . SiP.
By conversion : Some good man is an emperor . . . . PiS.

6. All cases of slavery are cases of leisure PaM ;
All cases of civilization are cases of leisure . . . . SaM ;
∴ All cases of civilization are cases of slavery . . . . SaP.
Fallacy of undistributed middle term.

7. All metals are conductors of electricity MaP ;
The atmosphere is not a metal . . . . SeM ;
∴ The atmosphere is not a conductor of electricity . . . . SeP

Fallacy of illicit distribution of the major term.

8. What is acquired without toil is not appreciated . . . . MeP;
The fortune of some people is acquired without toil. . . . . SaM;
∴ The fortune of some people is not appreciated . . . . SeP.

9. All visible bodies are material . . . MaP;
The rings of Saturn are visible . . . SaM;
∴ The rings of Saturn are material . SaP.

10. All material bodies are subject to the law of gravitation . . . . MaP;
The fixed stars are material bodies. . SaM;
∴ The fixed stars are subject to the law of gravitation . . . . SaP.

11. All voters are ten-pound householders. PaM;
Smith is a ten-pound householder . . SaM;
∴ Smith is a voter . . . . SaP.
Fallacy of undistributed middle term.

12. Most Italians are Catholics . . . . MiP;
X is an Italian . . . . SaM;
∴ X is a Catholic . . . . SaP.
Fallacy of undistributed middle term.

13. Most electors are in favour of female suffrage . . . . Most M's are P;
Most electors are Conservatives . . Most M's are S;
∴ Some Conservatives are in favour of female suffrage . . . . SiP.
Valid.

14. [No loyal member of a party criticizes its policy . . . . MeP ;]
X criticizes the policy of his party . SaP ;
∴ X is not a loyal member of his party SeP.

15. [No people who cannot co-operate with others achieve anything great . . . MeP ;]
Self-centred people are people who cannot co-operate with others . . . SaM ;
∴ No self-centred people achieve anything great . . . . SeP.

16. [What people suppose they know all about is a thing about which they never learn anything . . . . . . MaP ;]
Beauty is a thing which people suppose they know all about . . . . . . SaM ;
∴ Beauty is a thing about which people never learn anything . . . SaP.

17. [All plays which are romantic and have a happy ending are popular . . . . MaP ;]
Mr. B.'s new play is romantic and has a happy ending . . . . SaM ;
∴ Mr. B.'s new play is popular . . SaP.

18. [Those who treat their students as if they were mere buckets are not good teachers . . . . MeP ;]
Prof. Oracle treats his students as if they were buckets . . . . SaM ;
∴ Prof. Oracle is not a good teacher . SeP

19. Historians who are seen at their best are not those who cast their own shadows upon their pages . . . MeP ;

Macaulay, Mommsen and Treitschke cast their own shadows upon their pages . . . . SaM ;
[∴ M., M. and T. are not seen at their best . . . . SeP.]

20. [What can effect great permanent changes ought to be kept honest, etc. MaP ;]
Opinions can effect great permanent change . . . . SaM ;
∴ Opinion ought to be kept honest, etc. SaP.

21. [No person on whom cheerfulness always breaks in is a philosopher . MeP ;]
I am one on whom cheerfulness is always breaking in . . . . . . SaM ;
∴ I am no philosopher . . . . . . SeP.

22. Some coarse writers are popular . . PiM ;
The man of genius who catches the ear of the multitude is popular . . . SaM ;
∴ The man of genius, etc., is coarse . SaP.
Fallacy of undistributed middle term.

23. [All wealth has exchange value . . PaM ;]
Some valuable things have no exchange value . . . . SoM ;
∴ Some valuable things are not wealth SoP.

24. [No one who cannot keep a secret is trusted . . . . MeP ;]
Who loveth wine cannot keep a secret SaM ;
∴ Who loveth wine is not trusted . . SeP

25. [All radii of the same circle are equal . MaP ;]
AB and CD are radii of the same circle SaM ;
∴ AB and CD are equal . . . . SaP.

26. Wind and weather are unpredictable . MaP ;
Wind and weather are governed by laws MaS ;
[∴ Some things governed by laws are unpredictable . . . . SaP.]

27. Foresight is power . . . . MaP ;
Knowledge is foresight . . . . . . SaM ;
[∴ Knowledge is power . . . . . . SaP.]

28. [Nothing that casts suspicion on the judges is expedient . . . . . . MeP ;]
Every secret trial casts suspicion on the judges . . . . SaM ;
∴ No secret trial is expedient . . . SeP.

29. [Companions who ask no questions and pass no criticisms are agreeable . . MaP ;]
Animals ask no questions, etc. . . . . SaM ;
∴ Animals are agreeable companions . SaP.

30. Play is life. . . . . MaP ;
Interesting work is play . . . . SaM ;
[∴ Interesting work is life . . . . . SaP.]

31. [Those who have no capacity for enjoyment are incapable of making anybody happy . . . . MaP ;]
Some people have no capacity for enjoyment . . . . SiM ;
∴ Some people are incapable of making anybody happy . . . . SiP.

32. [People who express views about which commonplace people can say : " That is just what I always thought," are popular . . . . MaP ;]
Some authors express views, etc. . . . SiM ;
∴ Some authors are popular . . . SiP.

33. [What we look out for with special keenness we detect sooner than do others . . . . MaP ;]
What we really want we look out for, etc. . . . . SaM ;
∴ What we really want we detect sooner than do others . . . . . SaP.
Or in the corresponding pure hypothetical syllogism :
[If we look out for anything with special keenness, then we see it sooner than do others . . If B, then C ;]
If we really want a thing we look out for it with special keenness . If A, then B ;
∴ If we really want a thing we detect it sooner than do others . . If A, then C.

34. [If wishes were horses beggars would ride . . . . If A, then C ;]
But beggars do not ride . . . . . . not C ;
∴ Wishes are not horses . . . . . not A.

35. If all people were clever, there would be no examinations . . . If A, then C ;
[But there are examinations . . . not C ;
∴ Not all people are clever . . . not A.]

36. [If we lost the pleasure of hope, life would not be so interesting . If B, then C ;]
If we could see the future clearly, then we should lose the pleasure of hope
If A, then B ;
∴ If we could see the future clearly, then life would not be so interesting
If A, then C.

37. [If you only do what even the publicans do, then you have no merit . If B, then C ;]
If you only love them that love you, then you only do what even the publicans do . . . . . . If A, then B ;
∴ If you only love them that love you, then you have no merit . . If A, then C.

38. If the *Engravings* had been published in the sixteenth century, Anatomy would have been advanced by two centuries . . . . If A, then C ;
[But Anatomy was not advanced . . not C ;
∴ The *Engravings* were not published, etc. . . . . not A.]
Or the statement may have been intended merely as a detached (non-implicational) assertion.

39. If people are to bear having their follies ridiculed, they must be civilized If B, then C ;
[If there is to be comedy, then people must be able to bear having their follies ridiculed . . . . . If A, then B ;]
∴ If there is to be comedy, then people must be civilized . . . . If A, then C.

40. If the people before the Flood had no knowledge of the philosopher's stone, then Methuselah and others would not have lived so long. . . . If A, then C ;
But Methuselah and others did live long not C ;
∴ The people before the Flood had knowledge of the philosopher's stone not A.

41. If Darwin had not to write he would have been happy . . . . If A, then C ;
But he did write . . . . not A ;
∴ He was not happy . . . . . . not C.
Fallacy of denying the antecedent.

42. If Free Trade brought prosperity, then England would be the richest country If A, then C ;
England is the richest country . . . C ;
∴ Free Trade brings prosperity . . A.
Fallacy of affirming the consequent.

43. If all the absurd theories, etc., we should have no law and no religion left If A, then C ;
[But law and religion are still left in the world . . . . not C ;
∴ Absurd theories do not vitiate their objects . . . . not A.]

44. 1st sentence = If A, then C ;
2nd = not C ;
[∴ not A. The law is not really impartial, and does not punish blasphemy because it offends the feelings of believers.]

45. Pain is intolerable to the self-indulgent MaP ;
Restraint is pain . . . . SaM ;
[∴ Restraint is in tolerable to the self-indulgent . . . . SaP.]
Reserve is restraint . . . . RaS ;
[∴ Reserve is intolerable to the self-indulgent . . . . RaP.]

46. What fails to develop the power of independent judgment is no real education . . . . MeP;
The uncritical acceptance, etc., fails to develop the power of independent judgment . . . . SaM;
∴ The uncritical acceptance, etc., is no real education . . . . SeP.

47. Henry III of France was not moral . MeP;
Henry III of France was devout . . MaS;
∴ Some devout person was not moral SoP.

48. What raises the price of commodities raises the cost of living . . . . . MaP;
What keeps wages high raises the price of commodities . . . . SaM;
[∴ What keeps wages high raises the cost of living . . . . SaP.]
A shortage of labour keeps wages high RaS;
[∴ A shortage of labour raises the cost of living . . . . RaP.]
Indiscriminate doles mean a shortage of labour . . . . QaR;
∴ Indiscriminate doles raise the cost of living . . . . QaP.

49. To reduce the value of an examination is unjust to those who did well in it MaP;
To reduce the standard of an exam. is to reduce its value . . . . . . SaM;
[∴ To reduce the standard of an exam. is unjust to those who did well in it SaP.]
To be tender with weak candidates is to reduce the standard of the exam. . RaS;

∴ To be tender with weak candidates is to be unjust to those who did well RaP.

50. What makes life interesting . . . is beneficial . . . . MaP;
What makes the mind alert, etc., makes life interesting, etc. . . . . . . SaM;
[∴ What makes the mind alert, etc., is beneficial . . . . SaP.]
The struggle for existence makes the mind alert, etc. . . . . RaS;
∴ The struggle for existence is beneficial . . . . RaP.
[The actual conclusion is expressed more moderately by introducing "often."]

51. What increases trade cheapens articles, etc. . . . . MaP;
Free Trade increases trade . . . . . SaM;
[∴ Free Trade cheapens articles, etc. . SaP.]
What cheapens articles, etc., gives a greater purchasing power, etc. . . . PaR;
[∴ Free Trade gives a greater purchasing power, etc. . . . . SaR.]
What gives a greater purchasing power, etc., is a rise in real wages . . . RaT;
[∴ Free Trade is a rise in real wages . SaT.]
A rise in real wages is a boon to the working man . . . . TaV;
∴ Free Trade is a boon to the working man . . . . SaV.

52. Whatever shows that the same word may express different ideas, etc.,

helps one to discriminate between words and ideas . . . . MaP;

The knowledge of several languages shows that the same word, etc. . . SaM;

[∴ The knowledge of several languages enables one to discriminate between words and ideas . . . . SaP.]

To be able to discriminate between words and ideas is indispensable to clear thinking . . . . PaR;

[∴ The knowledge of several languages is indispensable to clear thinking . SaR.]

[What is indispensable to clear thinking is of great educational value . . RaT;]

∴ The knowledge of several languages is of great educational value . . . SaT.

[The transition from the second to the third syllogism is somewhat dubious —it is really assumed that a knowledge of several languages is the only help of the kind.]

53. Whosoever is not his own master is lacking in mental balance and sanity MaP;

He who has not learned obedience is one who is not his own master . . SaM;

[∴ He who has not learned obedience is lacking in mental balance and sanity . . . . SaP.]

He who is lacking in mental balance, etc., is incapable of an harmonious life . . . . PaR;

[∴ He who has not learned obedience is incapable of an harmonious life . SaR.]

[He who is incapable of an harmonious life is incapable of the highest life . RaT ;]

∴ He who has not learned obedience is incapable of the highest life . . . SaT

[The first sentence in the exercise only expresses this conclusion somewhat more abstractly.]

54. Where there is a lack of musical knowledge, there is a lack of musical curiosity . . . . MaP ;

Where there is an ignorant handling of music by unmusical editors, etc., there is a lack of musical knowledge SaM ;

[∴ Where there is an ignorant handling of music by unmusical editors, etc., there is a lack of musical curiosity . SaP.]

Where there is a lack of musical curiosity, people take no interest in their own greatest composers . . PaR ;

∴ Where there is an ignorant handling of music by unmusical editors, etc., people take no interest in their own greatest composers . . . . . SaR.

55. If there is an increased importation of gold, there is an increased issue of bank-notes . . . . If A, then C ;

If there is a favourable state of the exchanges, there is an increased importation of gold . . . . If B, then A ;

[∴ If there is a favourable state of the exchanges, there is an increased issue of bank-notes . . . . . . If B, then C.]

If there is an increased issue of bank-notes, there will be an advance in prices . . . . If C, then D ;

[∴ If there is a favourable state of the exchanges, there will be an advance in prices. . . . . If B, then D.]

If there is an advance in prices, it will check exportation and encourage importation. . . . . . . If D, then F ;

[∴ If there is a favourable state of the exchanges, it will (eventually) check exportation and encourage importation . . . . If B, then F.]

If there is a check on exportation and an increase of importation, it will tend to turn the exchanges against us . . . . If F, then G ;

[∴ If there is a favourable state of the exchanges, it will tend (eventually) to turn the exchanges against us If B, then G.]

[The events dealt with are represented as forming an economic cycle ; but the argument itself is not circular.]

56. A wealthy community is a productive one . . . . MaP ;

A happy community is a wealthy one . SaM ;

[∴ A happy community is a productive one . . . . SaP.]

[∴ The most happy community is the most productive community (Immed. Inf. by added determinants, DS is DP) . . . . DSaDP.]

The most productive community is that in which each citizen is allowed to enrich himself in his own way . DPaQ;

[∴ The happiest community is that in which each citizen is allowed to enrich himself, etc. . . . . . . DSaQ.]

The community in which each citizen is allowed to enrich himself in his own way is one in which there is profit-making in a free market . . QaR;

∴ The happiest community is one in which there is profit-making in a free market . . . . DSaR.

[The statement of the conclusion in the exercise is somewhat ambiguous—to say nothing about the way in which the argument ignores the distinction between the *production* and the *distribution* of wealth.]

57. If there is a reduction in the amount of money available for business, there will be a reduction in the number of workers employed If A, then C;

If there is high taxation, there is a reduction in the amount of money available for business . . . If B, then A;

∴ If there is high taxation, there is unemployment . . . . If B, then C;

If the Government has to spend a lot, then there is high taxation . If D, then B ;
If there is unemployment the Government has to spend a lot . . If C, then D ;
∴ If there is unemployment, then there is high taxation . . If C, then B.

58. If we have no reliable supply of raw materials, there can be no industry If A, then C ;
If we have no Colonies, we have no reliable supply of raw materials If B, then A ;
[∴ If we have no Colonies, there can be no industry . . . . . . If B, then C.]
If there is no industry, there is no prosperity . . . . If C, then D ;
[∴ If we have no Colonies, we have no prosperity . . . . If B, then D.]
∴ To have prosperity we must have Colonies . . . . . If not D, then not B

59. If there had been no German Navy, there would have been no Great War If A, then C ;
If there had been no Heligoland, there would have been no German Navy If B, then A ;
[∴ If there had been no Heligoland, there would have been no Great War If B, then C.]
If there had been no Great War, the fate of many millions would have been different . . . . . . If C, then D ;

[∴ If there had been no Heligoland, the fate of many millions would have been different . . . . . If B, then D.]

60. If nations have a sense of injustice, there is discontent . . . If A, then C;

If there is no juridical equality of all nations, they have a sense of injustice If B, then A;

[∴ If there is no juridical equality of all nations, there is discontent If B, then C.]

If there is discontent among nations, there can be no world-peace . If C, then D;

∴ If there is no juridical equality of all nations, there can be no world-peace . . . . If B, then D.

61. If there is no supply of labour, there can be no riches . . . . . . If A, then C;

If there is no poverty, there is no supply of labour . . . . If B, then A;

[∴ If there is no poverty, there can be no riches . . . . If B, then C.]

[If there is no wealth, there can be no civilization . . . . If C, then D;]

∴ If there is no poverty, there can be no civilization . . . . . . If B, then D.

[Note carefully the definition of "poverty" in the exercise.]

62. If there is to be adaptation to new conditions, there must be correction of old errors and the acquisition of new knowledge . . . . If A, then C;

If there is to be progress, there must be adaptation to new conditions If B, then A ;

[∴ If there is to be progress, there must be correction of old errors, etc. If B, then C.]

If there is to be correction of old errors, etc., there must be freedom of speech If C, then D ;

∴ If there is to be progress, there must be freedom of speech . If B, then D.

63. What requires a mental rearrangement is laborious . . . . MaP ;

The consideration of new ideas inconsistent with previous beliefs requires a mental rearrangement . . . . SaM ;

[∴ The consideration of new ideas, etc., is laborious . . . . SaP.]

What is laborious is condemned as an evil by the intellectually lazy . . PaQ ;

∴ The consideration of new ideas inconsistent with his previous beliefs is condemned as an evil by the intellectually lazy . . . . . . SaQ.

[As the average man is intellectually lazy, " average man " may be substituted for " intellectually lazy " in the conclusion ; but the actual conclusion in the exercise speaks of " new ideas " instead of " new ideas inconsistent with his previous beliefs," which is not legitimate, unless understood in this specific sense.]

64. The way to appear to be interested in others is to be really interested in them . . . . MaP;

The way to make friends is to appear to be interested in them . . . . SaM;

[∴ The way to make friends is to be really interested in them . . . . SaP.]

The way to be really interested in others is to cease to be self-centred . . . PaQ;

[∴ The way to make friends is to cease to be self-centred . . . . . . SaQ.]

65. If one has no courage, he has no manliness . . . . If A, then C;

If one has no strength, he has no courage . . . . If B, then A;

[∴ If one has no strength, he has no manliness . . . . If B, then C.]

If one has no manliness, he has no originality, except, etc. . . If C, then D;

[∴ If one has no strength, he has no originality, except, etc. . . If B, then D.]

66. This is a Simple Constructive Dilemma, with the minor premise omitted.

67. If one is a Jingo he likes *Abraham Lincoln* because of its patriotism; if one is not a Jingo, he likes it because of its broad humanitarianism.

[But everybody either is or is not a Jingo.]

∴ Everybody likes Abraham Lincoln.

Simple Constructive Dilemma.

68. If one watches the pictures of a film opera, he is disturbed by the non-synchronizing music; and if one listens to the music, he is disturbed by the non-synchronizing pictures.

[But one either attends to the pictures or to the music of the film opera.]

∴ One is disturbed in any case.

Simple Constructive Dilemma.

69. If Protection increases revenue, it cannot stimulate home industry; and if it stimulates home industry, it cannot increase revenue.

[But Protection either increases revenue, or it stimulates home industry.]

∴ Protection either cannot stimulate home industry, or it cannot increase revenue, i.e. it cannot do both.

Complex Constructive Dilemma.

70. If undergraduates attend Chapel because they care about religion, compulsion is unnecessary; and if they only attend Chapel because they are compelled, the compulsion is worthless in its results.

[But either they attend because they care about religion, or because they are compelled.]

∴ Compulsion is either unnecessary or worthless—absurd in either case.

Complex Constructive Dilemma, treated as Simple.

71. If Barrett did not tell the truth in regard to the solicitor, his evidence would be worthless; and if he did tell the truth, and actually had to be prompted by the solicitor for the defence, then his evidence would still be worthless.

[But Barrett either did or did not tell the truth.]

[∴ Barrett's evidence would be worthless in any case.]

Simple Constructive Dilemma.

72. Complex Constructive Dilemma treated as Simple in its main result. Minor premise omitted.

73. The light seen on parts of the moon not directly illuminated by the sun is either the moon's own light, or light reflected from the earth.

It is not the moon's own light.

∴ It is light reflected from the earth.

Mixed Disjunctive Syllogism.

74. If death deprives us of all feeling, it need not be feared on account of any alleged painful experiences it may bring; and if death gives us new kinds of sensations, it need not be feared on account of our alleged annihilation.

But death either deprives us of all feeling, or brings new kinds of sensations.

∴ Death need not be feared in either case.

Simple Constructive Dilemma. Its validity depends on the exhaustiveness of the possibilities stated, which is open to criticism.

75. (*a*) If this insect were a butterfly it would, when alighting, close its wings vertically.

[But it does not close its wings vertically.]

∴ This is not a butterfly.

Mixed Hypothetical Syllogism—Destructive.

(*b*) If this insect is a moth it will, when alighting, expand its wings horizontally.

It does expand its wings horizontally.

∴ It is a moth.

This Mixed Hypothetical Syllogism commits the fallacy of *Consequens*, unless the statement " moths expand them horizontally " is understood to mean that of these kinds of insects " *only* moths expand them horizontally," which would change the major premise into : " If this insect expands its wings horizontally when alighting, it is a moth," and the syllogism would be a Constructive Mixed Hypothetical Syllogism.

76 The main point of the argument is that the cancer experiments on mice cannot be defended on the ground that they are *both* useful in the study of the disease and painless to the mice —they are either not the one or not the other. The argument looks like a dilemma, but really consists of two hypothetical syllogisms.

(*a*) If the cancer is really like human cancer, it is not painless
If A, then not C ;
If the experiments are to be useful, the cancer must be like human cancer . . . . If B, then A ;
∴ If the experiments are useful, they are not painless If B, then not C.

(*b*) If the cancer is not like human cancer, the experiments are useless . . . . If not A, then not B ;
If the experiments are painless, the cancer is not like human cancer . . . . If C, then not A ;
∴ If the experiments are painless, they are useless . If C, then not B.

The two conclusions present two mutually exclusive alternatives—*either* useful but not painless, *or* painless but not useful.

77. (*a*) The violent and proud are pleased with power . . . . MaP ;
[Youth is violent and proud . . SaM ;]

∴ Youth is pleased with power . SaP.

(*b*) The placid and timorous are delighted with wealth . . . . . MaP;

[Age is placid and timorous . . SaM;]

∴ Age is delighted with wealth . SaP.

78. [What does not encourage independent thinking does not make thinking pupils . . . . MeP;]

Dogmatic teaching does not encourage independent thinking . . . . . . SaM;

∴ Dogmatic teaching does not make thinking pupils . . . . SeP.

79. [Those who have not sufficient opportunity for the exercise of their powers find self-government very difficult . . . . MaP;]

People accustomed to despotic rule have not sufficient opportunity for the exercise of their powers . . . SaM;

∴ People accustomed to despotic rule find self-government very difficult . SaP.

80. [What facilitates divorce loosens the marriage bond and weakens family life . . . . MaP;]

The Majority Report, etc., facilitates divorce . . . . SaM;

∴ The Majority Report, etc., loosens the marriage bond, etc. . . . . . SaP.

[What loosens the marriage bond, etc., is dangerous to the State . . . . . PaR;]

∴ The Majority Report, etc., is dangerous to the State . . . . SaR.

81. What is justifiable in the political domain is justifiable in the intellectual domain . . . . MaP;
Compromise is justifiable in the political domain . . . . SaM;
[∴ Compromise is justifiable in the intellectual domain . . . . . . SaP.
= Compromise in the intellectual domain is justifiable.]
The absence of complete candour from theological discussions is a kind of compromise in the intellectual domain. . . . . RaS;
∴ The absence of complete candour in theological discussions is justifiable . RaP.

82. [What leaves insufficient opportunity for the exercise of one's own will makes one weak-willed . . . . . MaP;]
Undue subordination to others leaves one insufficient opportunity, etc . SaM;
∴ Undue subordination to others (= excessive discipline) makes one weak-willed . . . . SaP.

83. If one is either perfectly wise, or utterly ignorant, he will not seek wisdom
If A or B, then C;
A philosopher does seek wisdom . . not C;
∴ A philosopher is neither perfectly wise nor utterly ignorant Neither A nor B.
i.e. something betwixt the two.
Destructive Mixed Hypothetical Syllogism.

84. If there are no warrants, there will be no dividends . . . . . . If A, then C;
If there are no casters, there will be no warrants . . . . If B, then A;
[∴ If there are no casters, there will be no dividends . . . . . . If B, then C.]
If there are no dividends, there will be no more loans . . . . . . If C, then D;
[∴ If there are no casters, there will be no more loans . . . . . . If B, then D.]
If there are no more loans, there will be no munitions . . . . . . If D, then F;
[∴ If there are no casters, there will be no munitions . . . . . . If B, then F.]
[But we must have munitions . . . not F.]
∴ We must retain our casters . . . not B.
Three pure hypothetical syllogisms, and one destructive mixed hypothetical syllogism—all abridged.

85. If industry is to exploit invention, each locality must share its productions with others . . . . . . If A, then C;
If the pressure of wants is to be met, there must be industry to exploit invention . . . . If B, then A;
∴ If the pressure of wants is to be met, each locality must share its productions with others . . . . . If B, then C.
But the pressure of wants must be met B;
∴ Each locality must share its productions with others . . . . C.
i.e. local barriers are burst by trade.

86. This is a straightforward Complex Constructive Dilemma.

87. Simple Constructive Dilemma with minor premise omitted. Can be construed as Complex.

88. If God is perfect, He is just, etc. If A, then C;
If God exists, He is perfect . . If B, then A;
[∴ If God exists, He is just, etc. If B, then C;]
If God is just, etc., my soul is immortal If C, then D;
[∴ If God exists, my soul is immortal If B, then D.]
If my soul is immortal, thirty years of life are nothing to me, etc. . If D, then F;
[∴ If God exists, thirty years of life are nothing to me, etc. . . If B, then F.]

89. If the inventor wants profits, there must be satisfaction of the consumer's interest . . . . If A, then C;
[The inventor does want profits . . A;
∴ There must be satisfaction of the consumer's interest . . . . . C.]
[Restate this conclusion so as to make its subject and predicate explicit—
The satisfaction of the consumer's interest is something which the inventor must satisfy . . . . . MaP;]
But the satisfaction of the consumer's interest is something which the inventor does not consciously aim at . MaS;

[∴ Something which the inventor does not consciously aim at is a thing which he must satisfy nevertheless . SiP.]

90. To have hopes is to be in the way of progress . . . . MaP ;

To have wants is to have hopes . . SaM ;

[∴ To have wants is to be in the way of progress . . . . SaP.]

∴ Sometimes not to have wants is not to be in the way of progress . . . $\bar{S}i\bar{P}$.

[Not to be in the way of progress is to be accursed . . . . $\bar{P}aR$;]

∴ Sometimes not to have wants is to be accursed . . . . $\bar{S}iR$.

91. If the Liberals are to help Mr. X to get into Parliament, he must change his views on the House of Lords ; and if the Conservatives are to help him, he must change his views on Free Trade . If A, then B ; if C, then D ;

[But he must obtain the help of either the Liberals or the Conservatives Either A or C ;]

∴ Mr. X will have to change his views either on the one or on the other point . . . . Either B or D.

The validity of the conclusion depends on the truth of the minor premise. The mere existence of other parties need not be fatal to it—if the other parties are either "Liberal" or "Conservative" in relation to these special problems. The chances of an "independent" candidate may be negligible

92. If the U.S. resolved to invade Canada, Canadian warships would be useless; if any other Power resolved to do so, Canadian warships would be unnecessary . . . If A, then B; if C, then D;
[But the (feared) invader must be either the U.S. or some other Power Either A or C;]
∴ Canadian warships would be either useless or unnecessary. . Either B or D.
[Expenditure on what is either useless or unnecessary is sheer waste of money . . . . MaP;
Expenditure on Canadian warships is expenditure on what is unnecessary or useless . . . . SaM;]
∴ Expenditure on Canadian warships is sheer waste of money . . . . SaP.

93. [Whatever teaches boys to suffer pain without resentment is a splendid training in self-control . . . . MaP;]
Our violent English games teach boys to suffer pain without resentment . SaM;
∴ Our violent English games are a splendid training in self-control . . SaP.

94. Societies in which there are no specialized functions are societies in which social intelligence is not fully developed . . . . MaP;

The peasant-state is a society in which there are no specialized functions . SaM ;

∴ The peasant-state is a society in which social intelligence is not fully developed . . . . SaP.

95. Many prominent lawyers make observations without any object . . . PiM ;

Sir John does not make observations without an object . . . . . . SeM ;

∴ Sir John is not a prominent lawyer. SeP.

Illicit process of the major term.

96. If an accused person is guilty, he ought not to be defended ; if he is innocent, it must be apparent to his judges [and he need not be defended]

If A, then B ; if not A, then C ;

The accused is either guilty or innocent

Either A, or not A ;

∴ The accused either should not, or need not be defended . . Either B or C.

[The actual conclusion is unwarranted. Why should not an " honest " lawyer defend an innocent person ? Moreover, is innocence always apparent to the judges ? May it not need the skill of another lawyer to make the innocence apparent ? Again, there are degrees of guilt. Is it safe to suppose that even the guilty will get no more than they deserve, if they are not helped to present their case effectively ?]

97. If the judges are not respected, the law will not be respected . If A, then C ;
[If the judges are not on their guard, they will not be respected . If B, then A ;
∴ If the judges are not on their guard, the law will not be respected If B, then C.
But the law ought to be respected . not C ;]
∴ The judges should be on their guard not B.

98. If it was wrong for Orestes, etc., Apollo should not have commanded him to do it ; if it was right, he should have protected him If A, then B ; if C, then D ;
[It was either wrong or right Either A or C ;
∴ Apollo should either not have commanded Orestes to do it, or he should have protected him . . . Either B or D.
As Apollo fulfilled neither alternative, he did not do what he should have done.]

99. What lowers self-respect, etc., is hostile to science and art . . . . MaP ;
Despotism lowers self-respect, etc. . . SaM ;
[∴ Despotism is hostile to science and art . . . . SaP.
What is hostile to science and art prevents cultural development . . PaR ;
∴ Despotism prevents cultural development . . . . SaR ;]
∴ The attainment of cultural development requires the removal of despotism (or the attainment of freedom) R̄aS̄.

100. Where education costs nothing, there the liberal professions are accessible to all classes . . . . MaP ;
In France education costs nothing . SaM ;
[∴ In France the liberal professions are accessible to all classes . . . SaP.]
Where the liberal professions are accessible to all classes, there is keen competition . . . . . . PaR ;
[∴ In France there is keen competition in the liberal professions . . SaR.
Where there is keen competition in the liberal professions fees are low . RaT ;]
∴ In France professional fees are low SaT.

GROUP J

THEORY OF INDUCTION

1.
2. } A. Wolf : *Essentials of Scientific Method*, Ch. I
3. W. E. Johnson : *Logic*, Vol. II, Ch. X.
4. W. S. Jevons : *Principles of Science*, Ch. VII.
5. H. W. B. Joseph : *Introduction to Logic*, Ch. XX.
6. J. S. Mill : *A System of Logic*, Book III, Ch. III.
   A. Wolf : *Essentials of Scientific Method*, Ch. VIII.
7. J. S. Mill : *A System of Logic*, Book III, Ch. V.
   A. Wolf : *Essentials of Scientific Method*, Chs. VII, VIII, IX.
8. B. Bosanquet : *Logic*, Vol. I, Ch. VI.
9. J. S. Mill : *A System of Logic*, Book III, Ch. X.
10. W. S. Jevons : *Principles of Science*, Chs. XVIII, XIX.
11. J. S. Mill : *A System of Logic*, Book V, Ch. IV.
12. W. S. Jevons : *Principles of Science*, Ch. XVIII.
    In the case cited much will depend on whether the ten witnesses were present at the time and place of the alleged theft, and whether the circumstances were such that the theft was unlikely to escape their notice.

13. W. S. Jevons: *Principles of Science*, Ch. XXIII.
16. A. Wolf: *Essentials of Scientific Method*, Ch. IX.
17. A. Wolf: *Essentials of Scientific Method*, Ch. II.
18. Those of Agreement, of Concomitant Variations, and of Residues.
19. A. Wolf: *Essentials of Scientific Method*, Ch. V.
20. In confining attention to "relevant" antecedents, when applying any of the methods of induction, we are treating the problem to some extent as if it were a residual problem.
23. A. Wolf: *Essentials of Scientific Method*, Ch. II.
24. 25. 26. A. Wolf: *Essentials of Scientific Method*, Ch. IX.
27. A. Wolf: *Essentials of Scientific Method*, Ch. VII.
28. J. S. Mill: *A System of Logic*, Book VI, Chs. VIII, IX.
29. Circumstantial Evidence.
W. S. Jevons: *Principles of Science*, Ch. XII.
30. A. Wolf: *Essentials of Scientific Method*, Ch. IV.
31. A. Wolf: *Essentials of Scientific Method*, Chs. III, VI.
32. By a study of correlations.
33. A. Wolf: *Essentials of Scientific Method*, Chs. VI, VIII.
34. A. Wolf: *Essentials of Scientific Method*, Ch. VIII.

35. A. Wolf: *Essentials of Scientific Method*, Ch. IV.

36. F. H. Bradley: *Principles of Logic*, Vol. I, Book II, Part II, Ch. III.

37.* J. Ward: *Naturalism and Agnosticism*, Lecture XIX.

38.* See No. 43 below.

39. A. Wolf: *Essentials of Scientific Method*, Ch. VII.

40. A. Wolf: *Essentials of Scientific Method*, Ch. IX.

42.* B. Bosanquet: *Implication and Linear Inference*, Ch. III.

43.* A. Wolf: *The Philosophy of Probability* (in Proceedings of the Aristotelian Society, Vol. XIII).

44.* A. Wolf: *Essentials of Scientific Method*, Ch. VI.

45.* J. N. Keynes: *Scope and Method of Political Economy*, Ch. II.

GROUP K

EXERCISES IN SCIENTIFIC METHOD

1. *Problem:* What is lightning?

   Hypothesis (suggested by *analogy*, or by certain similarities between lightning and electricity): Lightning is electricity passing from cloud to cloud.

   Verification: If the hypothesis is true a certain result should follow. Experiment accordingly, and the hypothesis is verified by the *Method of Difference* (in the presence of contact with the lightning conditions among the clouds the key when touched produces certain peculiar sensations which were not produced before such contact with the clouds).

2. *Problem:* How to exclude impurities from the lungs.

   Hypothesis: By the use of a cotton-wool respirator.

   Verification: When no such respirator has been used the air exhaled (and therefore the air inhaled) contains impurities; but after using a respirator no such impurities appear. *Experiment* by the *Method of Difference.*

3. *Hypothesis:* Dependence of the work of the

Royal Society in its early days on its first Secretary, Oldenburg.

Evidence: The Royal Society met regularly so long as Oldenburg was free to do his share; but did not meet at all during his long imprisonment. *Method of Difference.*

4. *Problem:* Relation between rust and air.

Hypothesis: Rust depends on that part of the air (oxygen) which supports flame and life.

Verification: The presence of moist iron filings in a given volume of air is followed by two concurrent changes—the iron filings rust and the volume of air is diminished. As no other relevant change has occurred these two facts must be connected. *Method of Difference.* Again, as the air in its previous state was such as could support flame and life, while the residue cannot, the part of the air used up must be the part capable of supporting flame and life. *Method of Residues.* [The use of this method would be more obvious if reference had been made explicitly to the fact that the iron filings, after rusting, gain just as much as the air loses in weight.]

5. *Problem:* The effect of ether on human beings.

Hypothesis: It affects their breathing.*

Verification: In each case tried the patient breathes more deeply after ether has been administered to him than before. *Method of Difference* in each case, the two instances being the two successive states of the same patient. When all the different patients are compared we get an illustration of the *Method of Agreement.* If one or two cases only had been tried, the result might have been attributed partly to constitutional peculiarities of the individual patients. The comparison of many different cases tends to eliminate this factor.

* An hypothesis may at first be comparatively vague—one may, e.g., suspect *that* ether affects one's breathing, without realizing *how.* The experimental verification may bring out the fact that it makes the breathing *more* rather than less deep, etc.

6. *Problem:* What causes these epidemics of colds?

Current Hypothesis: The arrival of ships.

Evidence: Probably many occasions on which the arrival of a ship was followed by such an epidemic. *Method of Difference* as regards each occasion, on the (erroneous) assumption that no other relevant change had occurred; the *Joint Method of Agreement and Difference* if many occasions (positive and negative) are compared.

Objection: No apparent connection between the alleged cause and the effect.

Auxiliary Hypothesis (to meet the objection, by inserting a connecting link): The effluvia given off by the new arrivals causes the epidemics. But this alleged agency is purely hypothetical; it is not a *vera causa.*

New Hypothesis: The north-east wind is the real cause of the colds. This is a *vera causa* of colds, and would therefore be accepted under the circumstances.

Evidence: Same methods as above, only substituting north-east wind for the arrival of a ship, which may be regarded as an irrelevant circumstance. The evidence might also have included the occurrence of epidemics after north-east wind when no ships arrived. Such evidence would be even more clinching; but we are not told.

7. *Problem:* The relation between human destiny and the heavenly bodies.

Old Hypothesis: A man's destiny is determined by the star under which he is born.

Evidence for: Not stated. Probably observation of a few coincidences according to the *Method of Simple Enumeration*, and the ignoring of exceptions.

Evidence against: If the hypothesis were true, then people born under the same star should have similar destinies; but they have not. *Method of Agreement with negative result.**

* The student is advised to make himself familiar with the conception of a *negative result* (not to be confused with a *negative instance!*). When an inductive method is applied to confirm or to test an hypothesis, and the result does not confirm the hypothesis, or even disproves it, then the result is said to be negative—it is not what it should have

been if the hypothesis had been true. Beginners usually find it difficult to realize that the method has the same essential character even if the result is negative—just as a train-journey remains a train-journey even if our friend happens to be out when we get there.

8. *Problem:* The relation between a string and the note produced by making it vibrate.

   Hypothesis: The length of the vibrating string determines the pitch of the note.

   Verification: The longer the vibrating string, the lower the pitch of the note produced, and vice versa. *Method of Concomitant Variations* (inverse concomitant variation).

9. *Problem:* Is there any connection between the intelligence of an animal and its nervous system?

   Hypothesis: There is.

   Evidence: The degree of adaptability of an animal's behaviour varies concomitantly with the complexity of its nervous system, and also with the degree of its intelligence. Therefore, the intelligence of an animal varies concomitantly with the complexity of its nervous system. The two are consequently connected with one another. Three applications of the *Method of Concomitant Variations.*

10. *Problem:* The nature of heat.

    Old Hypothesis: It is a kind of substance contained in material bodies (Phlogiston).

Evidence for: Not stated. But obviously based on such experiences as the emission of flame and heat by burning bodies.

Evidence against: If this hypothesis were true there should be some proportion between the size of bodies and the amount of heat they can hold. But there is no such proportion—two parts of the same body can be made to emit an almost endless amount of heat. *Method of Concomitant Variations with negative result.*

New Hypothesis: Heat is a form of energy.

Verification: The amount of heat produced is proportionate to the amount of mechanical work done. *Method of Concomitant Variations.*

11. *Problem:* What determines the colour of the yolk?

Hypothesis: The amount of green feed given to the hens.

Verification: The lot of hens which received no green feed laid eggs the yolks of which had no colour to speak of. The three lots which had green feed all laid eggs with coloured yolks. Group form of the *Method of Difference* (the three lots being regarded as one positive group). Again, among the three positive lots of hens the depth of the colour of the yolks varied concomitantly with the amount of green feed given to the hens that laid the eggs. *Method of Concomitant Variations.*

12. *Problem:* The influence of individual Cabinet Ministers on the government of the country.

Hypothesis: It is very slight.

Evidence: The country is governed in the same sort of way in spite of repeated changes in the Cabinet. But if the influence of Cabinet Ministers were great, there would be a difference after each change in the Cabinet. *Method of Difference with negative result*—a difference which makes no difference!

13. *Problem:* The influence of radium on sapphires.

Hypothesis: Radium may change the colours of sapphires.

Verification: After placing various sapphires under the influence of radium, without any other relevant change, the colour of each sapphire was changed at the end of a month. Four applications of the *Method of Difference,* the antecedent and subsequent state of each sapphire constituting the negative and the positive instance respectively in each case.

14. *Problem:* Why are grown-ups less interesting than when they were young?

Hypothesis: Education has spoiled them.

Evidence: Education appears to be the only relevant difference that has occurred in the interval. *Method of Difference.* Few people will probably agree to this if "education" is understood in the narrow sense of school and university teaching. So many other inner and outer changes take place in the interval, which are highly relevant. But if "education" is used in the wide sense, in which all our experience is part of our education, then the hypothesis has no special significance.

15. *Hypothesis:* Some of those who have been ostensibly punished for blasphemy have really been punished for being *uneducated.*

Evidence: If they had really been punished for *blasphemy*, then *all* writers of similar views should have been punished; but they were not—on the contrary some of their books are honoured with a place in the libraries of Bishops. *Method of Agreement with negative result.* That the punishment was really due to their *uneducated or crude way of expressing themselves* is shown by the fact that other writers, expressing essentially similar views, but not so crudely, have not been punished. *Method of Difference.*

16. *Problem:* Where is the seat of exertion when we move one of our limbs?

First Hypothesis: In the limb itself.

Evidence for: A feeling in the limb.

Evidence against: Though the arm remain intact it cannot move itself when the nerve connecting the limb with the brain or spine is severed anywhere. If the hypothesis were true this should constitute no essential difference. *Method of Agreement with negative result.*

Second Hypothesis: The exertion (or innervation) comes from the brain or the spinal cord.

Evidence: If the nerve connecting the limb with the brain or the spine is cut anywhere, the limb cannot be moved. *Method of Difference.*

17. *Problem :* What causes wort to deteriorate ?

Hypothesis : Dusty particles.

Verification : Wort protected against dusty particles remains uncontaminated ; but when dusty particles are allowed to enter it then the wort deteriorates. *Method of Difference.*

18. *Problem :* What are the conditions on which a high degree of civilization depends ?

Hypothesis : One * of the conditions is a certain type of climate [not specified].

Evidence : In all countries where a high degree of civilization is known to have existed in the past, or where it exists now, there the same type of climate existed or exists, though the countries may have been, or may be, different in many other respects. *Method of Agreement.*

* Generally speaking, all that the inductive methods usually show is that the alleged condition or agency is *one* of the conditions, not the *sole* condition. This has to be specially emphasized in some cases, but should always be remembered.

19. *Problem :* What is the function of the colours of the pupæ of butterflies ?

Hypothesis : To protect them (from birds, etc.) by making them indistinguishable from the surface on which they rest.

Evidence : The less conspicuous the pupæ are (or the more their colour is like the colour of the surfaces on which they rest) the more of them survive. *Method of Concomitant Variations.*

20. *Problem:* What is the cause of these different forms of butterfly?

Current Hypothesis: Differences in descent.

D.'s Hypothesis: Differences of temperature to which the chrysalis is submitted.

Evidence: Pupæ of the same descent, when subjected to various temperatures, evolve into various forms of butterfly. *Experiment* according to the *Method of Concomitant Variations.*

21. *Problem:* What is the relation between the size of towns and of the rivers which flow through them?

Hypothesis (obviously specious): The size of the town determines the size of the river.

Evidence: The two are found to vary together. *Method of Concomitant Variations.* This, however, could only prove that there is causal connection between them, not that the size of the town determines that of the river. The reverse is more likely to be the case, as the river was there first.

[Compare what happens when "canal," or "railway station," is substituted for "river," bearing in mind the possibility of reciprocal influence.]

22. *Problem:* What was the extent of Norse invasion of Yorkshire, Durham, and Northumberland?

Hypothesis: Northmen had invaded the coast south of the Tweed to an increasing extent, but had not got north of the Tweed at all, nor inland south of the Tweed.

Evidence: (*a*) The presence of Norse names on the coast south of the Tweed, and their absence inland and north of the Tweed. *Method of Difference.* (*b*) The diminishing frequency of Norse names on the coast north of Yorkshire. *Method of Concomitant Variations.*

23. *Problem:* Does the presence of copper influence the vibration of a magnetic needle?

Hypothesis: It does.

Verification: In the presence of a plate of copper a magnetic needle came to rest sooner than in the absence of the plate of copper, all other relevant conditions remaining the same. Therefore, the plate of copper must have exercised a retarding influence on the needle. *Method of Difference.* As the other retarding influences (the resistance of the air, and the imperfect mobility of the thread) could be allowed for, it was possible to estimate the retarding influence of the plate of copper alone. *Method of Residues.*

24. *Problem:* Why did the porters carry these straps?

Hypothesis: It was the survival of a habit acquired when the straps really were useful for carrying Sedan-chairs.

Evidence: No present purpose or function can

account for these straps. But men are known to be creatures of habit, and it can be shown that Sedan-chairs were in use at one time, and that such straps just served the purpose of carrying them. *Method of Residues.*

25. *Problem:* Do worms discriminate between leaves by the sense of taste or of touch?

Hypothesis: By the sense of taste.

Verification: Of two groups of leaves, each group as a whole containing leaves similar in texture to the leaves in the other group as a whole, but markedly different in taste, one is very much preferred by the worms. Therefore their choice must be guided by taste rather than by touch. The group form of the *Method of Difference*, or the *Joint Method*; the group of rejected leaves being regarded as a negative group relatively to the other group.

26. *Problem:* Relation between wealth and scale of production.

Hypothesis: Wealth is increased by an increase in the scale of production.

Evidence: A variety of cases in which an increase of wealth has followed an increase in the scale of production. Each case separately illustrates the *Method of Concomitant Variations*; all the cases together, the *Method of Agreement*, or, if account is taken of the negative instances vaguely referred to, the *Joint Method.*

27. *Problem:* Are the sensations which we derive from the muscles of a limb the principal factor in our appreciation of the direction and extent of the movement of the limb?

Hypothesis: Muscular sensations have no such influence.

Verification: If muscular sensations played an important part in our appreciation of the movement of a limb, then the absence of such muscular sensations, when the limb is moved in the way described, should prevent one from appreciating the various movements of the limb in the normal way. But that is not the case. *Method of Difference with negative result.* It is just the same as if the contradictory hypothesis had been tested and rejected.

28. *Problem:* The influence of climatic conditions on human fitness.

Hypothesis: Spring and autumn are the seasons most favourable to human fitness.

Evidence: The countries enumerated are different in various ways, yet in all of them it has been observed that the inhabitants show the greatest physical and mental fitness in spring and autumn. *Method of Agreement.*

29. *Problem:* Are rays of light ever bent out of their rectilinear path by the influence of gravitation?

Hypothesis: They are.

Verification: If they are, then rays passing near the sun should be bent towards the sun; and the star emitting those rays should consequently appear to occupy a somewhat different position then than usually. Generally speaking this is the *Method of Difference.* More accurately it is the *Method of Residues*—it is the residual difference between the usually observed position of the star in question, and its position when the sun is near (which can only be observed during an eclipse) that has to be accounted for by the hypothesis of Relativity.

30. *Problem:* What makes countries wealthy?

Hypothesis: Manufacture.

Evidence for it: All manufacturing countries are wealthy. *Method of Simple Enumeration,* or of *Agreement.*

Objection: The fact that manufacture and wealth are found together does not prove that manufacture is the cause of wealth (Fallacy of *cum hoc ergo propter hoc*). Otherwise you could prove that theatres are the cause of the wealth of cities. In support of this specious contention, evidence is cited following the *Methods of Agreement and of Concomitant Variations.* The moral intended is that manufacture is the result, not the cause of wealth, just as theatres are. The methods only show that certain things are causally connected, not which is the cause of which. But the objection overlooks the possibility of reciprocal influence between manufacture and wealth.

31. *Problem:* Why does wine exposed to air change to vinegar?

Liebig's Hypothesis: On account of the nitrogenous matter in the wine.

Evidence: Alcohol and water without nitrogenous matter does not turn into vinegar, but when nitrogenous matter is added it does change to vinegar. *Method of Difference.*

Pasteur's Hypothesis: The change is caused by the mycoderm.

Evidence: Alcohol and water without anything capable of supporting mycoderm organisms does not acetify; but when saline crystals (*not* nitrogenous matter) are added acetification takes place. *Method of Difference.* The absence of nitrogenous matter in Pasteur's experiment disproves Liebig's hypothesis (*Method of Difference with negative result*). The nitrogenous matter in Liebig's experiment only served as a medium for mycoderm organisms.

32. *Hypothesis:* Bathing in the Serpentine (or in the open) every morning throughout the year is healthy.

Evidence for: Here is a person who has bathed in the Serpentine for forty years and is alive and well.

Objection: Where are the others who once bathed there? Are they all well, or have any of them become ill in consequence? One case is not sufficient evidence. It may be the fallacy *cum hoc ergo propter hoc.* He may be alive and well in spite of, rather than because of, the daily dip. Or what may suit his robust constitution, may be fatal to a less robust constitution. The *Method of Simple Enumeration* is never strong, and is particularly unreliable when negative cases have been ignored.

33. This illustrates the formation of an *Hypothesis by Analogy*. To make it more than a mere hypothesis, more evidence will be required—namely, the kind of evidence sufficient to establish the law of refraction and its application to the atmosphere.

34. *Problem:* What is the cause of the immense geological changes?

Earlier Hypothesis: Sudden catastrophes.

Lyell's New Hypothesis: The ordinary slow causes of land-corrosion, etc. Both hypotheses relate to causes (or agencies). The new hypothesis has the advantage that it suggests a *vera causa* (an agency of the existence and operation of which there is independent evidence); of the "sudden catastrophes" of the earlier hypothesis there is no independent evidence. Hence the new hypothesis is preferable.

35. *Problem:* To account for the presence of shells found at a great height above the sea.

Hypotheses: There are five rival hypotheses, each suggesting a different cause or agent. Now hypotheses with *veræ causæ* (causes otherwise known to exist) are preferred to those with purely hypothetical causes. This disposes of the first two hypotheses. Again, of *veræ causæ* preference is given to those which are known not only to exist, but also to act in the kind of way required. This disposes of the third hypothesis. The fourth hypothesis cannot account for *all* the shells, and would therefore still leave a residual problem. The last hypothesis is the only one that would account for all the facts, and it is supported by close analogies of observed facts.

36. This illustrates the *Deductive-Inductive Method.* By deduction from the hypothesis or law of gravitation Newton arrived at the conclusion stated. Subsequent inductive investigations confirmed it.

37. *Problem:* Relation of commercial prosperity to armaments.

Hypothesis of Mr. H.: Commercial prosperity depends upon armaments.

Evidence pro: [Not stated. Presumably the concurrence of the two in some cases. *Method of Simple Enumeration,* or of *Agreement.*]

Evidence contra: The small Powers who are commercially as prosperous as their powerful neighbours constitute apparent exceptions to the hypothesis, and must be fatal to it unless specially explained.

38. *Problem:* Whence come these "thunderbolts"?

Terrestrial Hypothesis: From terrestrial volcanoes.

Lunar Hypothesis: From lunar volcanoes. This hypothesis does not deal with a *vera causa.* Moreover, it does not account for the velocity of the "thunderbolts"; less still can the terrestrial hypothesis account for it. Neither hypothesis, moreover, can account for the direction of the motion of the "thunderbolts."

Cosmic Hypothesis: Chladni's hypothesis is the only one which, if true, would account for all the relevant facts, and is accordingly accepted—subject always to revision if new evidence should call for it.

39. *Problem:* What determines workmen's wages?

Hypothesis: The minimum necessary to keep them alive ("the iron law of wages").

Evidence for: Not stated. Probably the fact that so many workers are near the poverty line.

Objections: The various exceptions cited. Real exceptions are fatal to an hypothesis. The upholder of the hypothesis must therefore show special reasons such as would make the exceptions only apparent, not real. In each case such a special reason for the exception is suggested, but only by way of derision.

40. *Problem:* What is the cause of the putrefaction of organic matter?

Hypothesis: Floating particles in the air.

Evidence: Where there are no such floating particles in the air, no putrefaction occurs; where there are such floating particles, putrefaction does occur. *Method of Difference.* The more numerous such floating particles in the air, the more extensively and the more rapidly does putrefaction take place in exposed organic matter. *Method of Concomitant Variations.*

41. A simple illustration of the *Deductive-Inductive Method.* From Pasteur's hypothesis Lister deduced certain consequences bearing on surgery, and confirmed those deductions inductively, by the *Method of Difference.* In his ward, where the antiseptic treatment was used, there was practically no gangrene; while in neighbouring wards, in which the antiseptic treatment was not in use, cases of gangrene were common.

42. *Problem:* The nature of animal tissue.

Hypothesis: It is cell-like in structure.

Evidence: Similarity in superficial appearance to the more obviously cell-like structure of vegetable tissue. A kind of analogy.

Verification: If the resemblance of animal to vegetable tissue is not merely superficial, but essential, then, since the nucleus was known to be so important in the vegetable cell, the cell in animal tissue should also have a nucleus, though no such nucleus had till then been observed. Subsequent microscopic investigation showed this to be the case, and so confirmed the cell hypothesis.

Here we see how analogy may *suggest* an hypothesis; the hypothesis, however, is not established by the mere analogy, but by the subsequent inductive investigation—in this case by actual observation by means of the microscope. The generalization will depend on the Principle of Fair Samples only, not on any of the "inductive methods."

43. *Problem:* Why should clover be more abundant in fields near villages than in others?

Evidence: In a number of fields near villages the clover is more abundant than in outlying fields. The *Joint Method of Agreement and Difference* suggests a probable connection between the abundance of the clover and the nearness of the fields to the villages. But the connection does not look plausible, looks rather remote. Accordingly, Darwin sought and discovered an *auxiliary hypothesis* relating to the connecting-links between the apparently remote facts. As the general relationship between clover-flowers, bees, mice, and cats was already known, the auxiliary hypothesis can also be treated as a deductive application of already established truths to this class of case. In that event we have here a deductive confirmation of an induction based on the Joint Method, and the whole investigation is an instance of the *Inductive-Deductive Method* (Mill's *Historical Method*).

44. *Problem:* To account for the difference between the recorded and the calculated dates of ancient eclipses.

It is not stated whether the calculated date

falls earlier or later than the recorded date; but the character of the solution should make this clear.

Hypothesis: The day is now longer than it used to be (or the day used to be shorter than it is now). This means that in the past there were more days in a longer period, such as a month, than there are now. For a day is simply the time taken by the earth to rotate once round its axis, while a month is the time taken for the moon to revolve round its orbit. If in the past the moon revolved with the same speed as now, and the earth rotated more rapidly, then there would have been more than the present number of days to the lunar month; and so with other longer periods. The effect of the hypothesis would be to equate the number of days, between now and the ancient eclipses, with a *smaller* number of months, years, etc., than would otherwise be the case. [So that the calculated date must have gone too far back.] *Method of Residues.* But why should the days be longer now than they used to be?

Auxiliary Hypothesis: Effect of the tide on the earth's rotation. This is the connecting-link. Compare No. 43.

45. This illustrates the difficulty and uncertainty of associating any result with a single change in a highly complex whole. The case of England would suggest that Free Trade is

the cause of prosperity; that of Germany and the United States would suggest that Protection is the cause of their prosperity. The real cause of prosperity in either case is a whole complex of conditions in which Free Trade or Protection may have played a minimal part, or in which these opposite trade policies may have helped to produce similar results because each was combined with different other conditions. The chief fallacy to be guarded against in all such cases is that of arguing *post hoc ergo propter hoc.*

46. *Problem:* How do different kinds of birds come by their different kinds of song?
Current Hypothesis: By heredity.
New Hypothesis: By imitation.
Evidence: (1) For heredity. Not stated Probably common coincidence of differences of descent and differences of song. *Method of Agreement* or of Simple Enumeration.
(2) For imitation. Linnets when shut up with different kinds of larks sing none of them like a linnet brought up by another linnet, but each like the lark with which it had been shut up. *Experiments* by the *Method of Difference.*
There is no *pure* negative instance. (A young linnet brought up with no other bird would perish.) But each instance can be treated in turn as negative relatively to the others.
The final conclusion is much too wide, as the experiments were confined to linnets.

47. *Problems:* (1) Has the crayfish a sense of smell?

(2) If so, where is its organ of smell?

Hypothesis (1): It has.

Evidence: When odours are introduced (without any other relevant change being effected) it retreats. *Method of Difference.*

Hypothesis (2): The organ of smell is located in the antennules.

Evidence: Of two otherwise similar crayfish, the one with antennules retreats from odours, while the one without antennules does not. Hence, by the *Method of Difference,* the antennules are a necessary condition of sensibility to odours.

48. *Problem:* In what way does the result of drinking raw spirits differ from that of drinking matured spirits?

Hypothesis: Raw spirits produce a more violent condition.

Verification: Both monkeys become violent after drinking raw spirits, but only foolish after matured spirits. Group form of the *Method of Difference,* the cases of intoxication with matured spirits being treated as a negative instance relatively to the cases of intoxication with raw spirits.

The final conclusion rests on the assumption that what is true of these monkeys will be true not only of all monkeys but of animals generally.

49. *Problem:* Do all acids contain oxygen?

Old Hypothesis: They do.

Verification: Ammonia is known not to contain oxygen. If water should be produced on combining ammonia with hydrochloric acid, then the oxygen required for the water (H_2O) could only be accounted for as derived from the acid. *Method of Residues.* But the result is *negative*; the hypothesis is *not* confirmed. Such dew as was formed could be accounted for without the aid of the acid. [Might also be construed as Method of Difference, of which the Method of Residues is a special form.]

50. *Problem:* What happens when wine changes to vinegar?

Hypothesis: The wine is oxidized—takes up oxygen from the air.

Verification: Part of the bottle contained wine, and all the rest of the bottle contained air at the beginning of the experiment. At this stage no water would have entered the bottle when submerged in the way described. But when the wine had changed into vinegar, enough water rushed into the submerged bottle to fill one-fifth of the space originally filled by air. *Method of Difference.* Now that was the volume of oxygen originally in the bottle; and the remaining air had the properties of nitrogen, not of oxygen. *Method of Residues.*

51. This is an example of the *Method of Residues.* Allowing only for the movements of the

planets *within* the solar system, the position of the sun among the stars should have been approximately the same in the time of Ptolemy (allowing for errors of observation). But it is very different. What is the cause of the residual difference? The only hypothesis that seems to account for it is the hypothesis that the whole solar system has itself changed its position in space; and if it has done so in the past, it is most probably also doing so now.

52. A striking illustration of the *Evolutionary Method*. From his study of fossil remains, representing important stages in the evolution of fishes, Agassiz obtained a sufficiently definite idea of their whole evolution to enable him, by the scientific use of his imagination, to interpolate intermediate links.

53. *Problem:* What causes grape-juice to ferment?
Hypothesis: Minute organisms entering the grape-juice from outside.
Verification: The juice taken from a sound grape and from which dust, etc., had been excluded (1st and 4th series of flasks) did not ferment; similar juice, to which had been added some washing water containing dust collected from the surface of grapes and stalks (2nd series of flasks), did ferment. The cause of the fermentation is therefore something external to the grape-juice itself. *Method of Difference*, group form, as there were 20 flasks in one instance, and 10 in

the other. Again, no fermentation occurred when the washing water containing the dust was boiled before being added to the grape-juice (3rd series); whereas it did occur when the washing water was not boiled first (2nd series). The boiling of the water therefore made a real difference. *Method of Difference*, group form. But this fact can be best accounted for on the auxiliary hypothesis that what really caused the fermentation was the kind of living organism which had actually been observed in the washing water (*vera causa*), and which would be destroyed by boiling.

54. *Problem:* To account for the sudden quickening of the patient's pulse.

Hypothesis: It was caused by the mention of the name of Pylades.

Evidence: It followed immediately after the mention of that name. *Method of Difference.* But this might be a fallacious inference (*post hoc*); something else may have caused it. Hence need for verification.

Verification: Under exactly similar circumstances there was no quickening of the pulse when other names were mentioned; but there was, when Pylades' name was mentioned. Two positive and two negative instances altogether. Really several applications of the simple *Method of Difference*, taking one positive and one negative instance at a time. But it can also be regarded as an instance of the Joint Method of Agreement and Difference.

The strictly correct conclusion is that the mention of Pylades' name caused the quickening of the lady's pulse. That she was in love with him is an auxiliary hypothesis (linking up the name with the pulse) which may not be correct—the same might have happened if she hated him, etc.

55. Two applications of the *Method of Difference* show that light is a condition of the formation of starch in plants. The final conclusion is an additional (or auxiliary) hypothesis to explain how the light acts in the formation of the starch, namely, by decomposing the carbon-dioxide gas in chlorophyl.

56. This is an application of the *Method of Concomitant Variations* to the two series of phenomena. But no additional or auxiliary hypothesis is offered relating to the link or links between the two series. Contrast Nos. 43, 55, etc.

57. This is a straightforward illustration of *the group form of the Method of Difference.*

58. *Problem:* The origin of the minute organisms found in water in which animal or vegetable matter had been decomposed.

Rival Hypotheses: (1) Germinal origin; (2) spontaneous generation.

Verification: If living germs were necessary to originate such minute organisms, then they should not appear where all germs had been destroyed. Now Needham believed that by heating the animal or vegetable matter and the water he destroyed all the germs they might contain; and as the jars were hermetically sealed, no new germs could enter. He held, accordingly, that, if the hypothesis that living germs are necessary to originate such organisms were true, no organisms should be found in these jars. But they were. *Method of Difference* (negative instance) *with negative* (or contradictory) *result.* He concluded, therefore, that such organisms can originate spontaneously without living germs.

Spallanzani's experiment showed, however, that Needham had not really eliminated all living germs by his process;[1] and that when all living germs are effectively destroyed, then no new organisms come into being. In other words, Needham's alleged negative instance was not negative but positive. Spallanzani supplied a real negative instance, and his conclusion rests on the *Method of Difference.*

59. *Problem:* To discover the cause of the spread of yellow fever.

Hypothesis: Mosquitoes spread the disease. Suggested by analogy (malaria).

Verification: First experiment—Dr. L., who was well before he was bitten by a mosquito, contracted yellow fever soon after he was bitten. *Method of Difference.*

[1] Note the importance of employing reliable *technical* methods. Needham's faulty *technique* betrayed his *logic.*

The second experiment was to test the question whether other things than mosquitoes could communicate yellow fever after they had been in contact with patients suffering from it. The result was negative.

The third experiment shows two spaces alike in all relevant respects except that mosquitoes are present in one of them, but not in the other. The volunteer in the part where the mosquito was, contracted the fever; the others did not. *Method of Difference.*

60. *Problem:* To determine what part of the brain is connected with our memory of words.

Hypothesis: A certain convolution of the left frontal lobe of the brain.

Evidence: Comparing the brain of a normal person (after death, of course) with that of a man who had lost the power of speech, it appeared to Broca that the only relevant difference between them was that the convolution in question was sound in the former case, but destroyed in the latter case. *Method of Difference.*

61. *Problem:* What is the cause of the immunity of modern communities from smallpox?

Current Hypothesis: The practice of vaccination.

Evidence (implied): the introduction of general vaccination was followed by comparative immunity from smallpox. *Method of Difference.*

Objection: This may be a fallacy (*post hoc ergo propter hoc*). In Iceland such periods of immunity occurred without vaccination.

[The passage leaves the problem open. The claims of vaccination really rest on comparative statistics of immunity among the vaccinated and the unvaccinated during epidemics.]

62. *Problem :* What determines national differences ?

Old Hypothesis : Climatic and other physical differences of habitat.

Evidence for : Not stated. Probably obvious coincidence of both kinds of differences in certain cases. *Method of Simple Enumeration.*

Evidence against : (1) The cases cited in which we find national *differences* where the climatic and other physical conditions are *similar ;* and (2) cases alleged (but not cited) in which peoples are *similar* although their climatic and other physical conditions are *different.* The first cases (1) illustrate the *Method of Agreement with a negative result* (if the hypothesis were true similar climatic conditions should always result in similar nationalities, but they don't). The second cases (2) illustrate the *Method of Difference with a negative result* (if the hypothesis were true, then different climatic conditions should always result in national differences, but they don't).

The final conclusion would be more correct if the last words were " not the *only* forces that make nations."

63. *Problem :* Are all the different rays composing white light detrimental to milk ?

Hypothesis: Some are and some are not.

Verification: The milk in red glass bottles did not "turn," while the milk in bottles permitting other than red rays to pass through did "turn." Therefore red rays are not injurious, while all others are. *Method of Difference,* treating the red bottles as negative instances relatively to each of the others. Again, the milk in the bottles which are not red had "turned" more or less according as the colour of the rays passing through the glass bottles was nearer to or farther from the violet end of the spectrum. *Method of Concomitant Variations.*

64. *Problem:* To determine the value of certain fertilizers.

Experimental Test: Plots A and B are treated alike, except that B received these additional fertilizers. The fertility of plot B (as measured by the value of the milk given by the cows that grazed on it) was greater than that of plot A. *Method of Difference.* This requires that the two instances shall be as like as possible except in respect of the difference studied. To allow for possible differences in the cows, these were changed over fortnightly. But there still remained the possibility of an initial difference in the natural fertility of the plots. Hence the reversal of the experiment in 1913. *Method of Difference* again, with plot B as negative instance this time. Plot A now received the extra dressing. As it was more productive, the value of the fertilizers is confirmed. [Would it have been equally conclusive if one plot only had been treated in these different ways in two successive periods, or if both plots had been treated similarly *at the same time*? If not, why not?]

65. *Method of Concomitant Variations.*
66. *Method of Concomitant Variations* (between different degrees of light and of activity on the part of the leaf), and *Method of Difference* (when an instance of the presence of light is compared with its total absence).
67. *Method of Agreement.*
68. *Problem:* What causes modification of species?

Gulick's Hypothesis: Isolation itself.

Evidence for: Not stated. Presumably cases in which isolated species changed without obvious external causes. *Method of Simple Enumeration.*

Objection: If isolation itself were a real cause of change, then all isolated species should show change. In that case Ireland would be an exception fatal to the hypothesis. *Method of Agreement with negative result.*

Second Hypothesis: Natural selection by the struggle for existence causes modification of species.

Evidence for: Not stated, but presumably similar to that above.

Evidence against: Why no change in Irish species, which must surely have had to struggle for their existence? Again *Method of Agreement with negative result.*

Retort: The Irish species may not be a real exception. Possibly, owing to the small number of competing species there was no real struggle for existence.

[Note: In this passage the term *vera causa* is used in the loose sense of a *real cause,* not in the correct sense of *an independently known cause.*]

69. *Problem:* What is the nature of the process of nitrification?

Old Hypothesis: It is entirely chemical.

Evidence: Not stated, but probably based on the analogy of similar changes.

New Hypothesis: Nitrification is an organic process effected by bacteria. The old hypothesis did not account for the initial delay of twenty-one days; the new one does.

Verification: Under otherwise essentially similar circumstances nitrification proceeded when there was no chloroform vapour, but ceased when chloroform vapour was added. By the *Method of Difference* we conclude accordingly that chloroform vapour is a hindrance to nitrification. But this is only explicable on the hypothesis that the process of nitrification is brought about by living organisms (bacteria), which alone would be affected by chloroform vapour.

70. Here we have an *Hypothesis* which cannot be tested directly, but only indirectly, namely,

by deducing consequences from it until we arrive at such consequences which can be tested by observation. In this case the consequences are such as observation does *not* confirm. The hypothesis must, therefore, be abandoned.

71. An example of the *Method of Difference.*

72. *Problem:* To discover a remedy for rabies and hydrophobia.

Series of Hypotheses and Verifications: Following the analogy of vaccination against small-pox, Pasteur has to discover first a means of communicating the disease. This he discovered by the *Method of Difference*—by injecting a healthy animal with the saliva, etc., of an infected one. Next it was necessary to discover a way of varying the strength of the virus. This he discovered by submitting the virus to varying temperatures and noting the variations in the violence of the disease communicated by the corresponding injections—*Method of Concomitant Variations.* He next showed that dogs first vaccinated with weaker injections survived strong injections which other dogs did not survive. Group form of the *Method of Difference.* This showed that the preliminary vaccination secured immunity. Then, guided by analogy, he inoculated similarly victims of hydrophobia, and found the mortality rate among patients so treated was considerably less than among those not so treated. This showed a correlation between such vaccination and immunity, by the *Statistical Method of Correlation.*

73. *Problem:* To determine the law of falling bodies.

First Hypothesis: The speed is proportional to the *distance* fallen through. But this supposition involved self-contradiction.

Second Hypothesis: The speed is proportional to the *time* of the fall.

Verification: *Direct* verification being impracticable he tried *indirect* verification, namely, he deduced the consequences of the hypothesis until he found one which could be tested experimentally. The consequence deduced in this case was, that the *distance* through which a body (falling with a velocity proportional to the time) should fall would be proportional to the *square of the time.* This he could test, in the way described, according to the *Method of Concomitant Variations.* An *exact quantitative law* has this advantage that if it fits the facts no other law will.

[Note the difference between an *hypothesis of law* and an *hypothesis of cause*, exemplified respectively in this and in the preceding exercises.]

74. *Problem:* Is air a chemical combination or only a mechanical mixture of nitrogen and oxygen?

Hypothesis: It is only a mechanical mixture.

Verification: When air is confined in a vessel with a sieve, as described, it is found that the air drawn through the sieve contains more than the normal proportion of nitrogen. Now on the supposition of a chemical combination we could account for the normal proportion of nitrogen, but *not for more;* the difference would then remain an unexplained residuum. This *residuum,* however, can be accounted for by the present hypothesis, namely, that the nitrogen particles and the oxygen particles are only intermingled and can move independently of one another, so that under the special circumstances described more of the nitrogen particles than of the oxygen particles will pass through the sieve. *Method of Residues.* (In so far as the rival "chemical combination" hypothesis is refuted we may see also the *Method of Agreement with a negative result*—air with the normal proportion of nitrogen and oxygen on one side of the sieve leads one to expect the same proportion on the other side, but the proportion is not the same.)

75. *Problem:* Does woman's suffrage effect a reduction in infant mortality?

Dr. T.'s Hypothesis: It does.

Evidence for: In Australia and New Zealand the granting of woman's suffrage was followed by a reduction in the infant mortality rates. *Method of Difference.* This would be a correct inference if no other relevant change had taken place. But that is difficult to maintain.

Evidence against: In England and Wales there was also a reduction in the infant mortality rate during the same period, notwithstanding the absence of woman's suffrage. Other causes must have been operative here, and the same causes may have been operative in Australia, etc., woman's suffrage being only a coincidence.

The whole can be regarded as an illustration of the *Statistical Method of Correlation with a negative result.* To establish a correlation between woman's suffrage and a reduction in infant mortality, it would be necessary to show a much greater reduction where there is a woman's suffrage than anywhere else. But that is not so.

76. *Problem:* To explain certain peculiarities in the growth of peas and allied plants as compared with cereals.

Data (or evidence): The growth of cereals is proportionate to the amount of nitrate supplied as chemical fertilizer. *Method of Concomitant Variations.* In the case of peas, there is no such proportion—*Method of Concomitant Variations with negative result.* The successful peas, unlike the cereals, contained *more* nitrogen than had been given to them as chemical manure. This extra amount of nitrogen, therefore, constitutes an unexplained residue—a problem for the *Method of Residues.*

Hypothesis: The growth of peas and allied plants is effected by bacteria. This hypothesis, if true, would account for the extra amount of nitrogen in peas—*Method of Residues.*

Verification: Under otherwise similar conditions peas grow well when they could associate with bacteria, but do badly when bacteria are excluded. Therefore bacteria are active agents in the growth of peas. Group form of the *Method of Difference.* The agency suggested is a *vera causa* as there is independent evidence of its existence and of its absorption of nitrogen from the air.

77. This is a warning against the fallacious argument *post hoc ergo propter hoc,* which is so readily committed when, by a too easy-going application of the *Method of Difference,* two successive states of a country, etc., are treated as essentially alike except for one antecedent and one consequent, other relevant differences, possibly much more relevant differences, being ignored. The picturesque antecedent to which the consequent is entirely attributed in such cases may be little more than one condition, or merely an occasion of the consequent, possibly not even that; while the real forces which have produced the result in question may have to be sought among the ignored or overlooked antecedents.

78. *Problem:* Are the reactions of animals and of plants to light identical?

Loeb's Hypothesis: They are.

Evidence for: In both alike the same kind of external light stimulus is followed by, or causes, a turning towards the light. *Method of Difference.*

Objection: In this case induction takes the form of extending an accepted conception to a class of phenomena not previously classed under it. This is sometimes justified, as e.g. when lightning was included in the class of electrical phenomena. But it is not legitimate unless there is essential similarity, or identity, between the two kinds of phenomena hitherto regarded as distinct. In the present case, though there are external similarities as regards stimulus and ultimate change of position, yet *the internal processes by which the change of position is effected are totally different* in the case of animals and of plants, and should not be classed together without further ado.

79. Here we have a threefold statistical application of the *Method of Concomitant Variations.*

80. *Problem:* What is the function of the fiery spark of the firefly?

Current Hypothesis: To warn birds, insects, etc., that it is uneatable.

Objections: (1) Many fireflies are diurnal; and, since the fiery spark does not shine by day, it cannot in these cases function in the way supposed. (2) Most insect-eating birds are diurnal, and, for the same reason, cannot therefore be warned in the way alleged. (3) Raptorial insects feed freely on fireflies, and so are obviously not warned by the fiery spark in the way suggested. (4) Moreover, in the case of other animals, fire does not *warn* but *confuse* them, so that the hypothesis has no support even in analogy.

At all events, the exceptions (1–3) call for some serious modification of the hypothesis, if it is to fit the cases.

81. *Problem:* What is the antidote to scurvy?

Old Hypothesis: Vegetable acids.

Evidence: Chiefly the marked diminution in the amount of scurvy among voyagers after the general use of lime-juice on ships. *Method of Difference.* This would be conclusive if no other relevant change had taken place at the same time. But other changes did take place, notably, improvement in food, and reduction in the length of voyages. Both these were relevant factors, in the light of subsequent investigations (details of which are not given) which tend to show that scurvy is a form of poisoning caused by decayed meat. Reduction in the opportunities of eating decayed meat (by improving the food and shortening the voyages) would thus account for the reduction in the amount of scurvy. It is one of the dangers of the *Method of Difference* that some relevant change may be overlooked, even when the change which is stressed is also of importance. In this case the lime-juice may have helped, even if its importance was exaggerated.

82. This exercise is essentially the same as No. 62, with one important addition. It is argued

that the hypothesis receives no support even from analogy. Even in the case of the lower animals physical conditions are not the sole or most important determining conditions of the character of species. In Borneo and New Guinea the physical conditions are essentially similar, yet the species of animals are very different. *Method of Agreement with negative result.* On the other hand, in Australia and New Guinea, the physical conditions are different, while the animal species are similar. *Method of Difference with negative result.*

83. This is a warning against the careless application of the *Method of Difference.* Instead of adding the one antecedent or factor which is under investigation, others are (it may be unwittingly) added at the same time, and the resulting change is attributed solely to the one antecedent. This is a common source of the fallacy *post hoc ergo propter hoc.* Compare Nos. 77, 81, etc.

84. *Problem:* To explain the immunity of birds from anthrax.

Hypothesis: It is due to their warm blood.

Evidence: *Partly deduction* from the general truth already established that anthrax microbes do not develop at a temperature of 44° C. upwards. *Partly induction*—an inductive confirmation of the deduction. The whole is an example of the *Deductive-Inductive Method.* The inductive part follows the

Method of Difference—when the bird's blood is warm it is immune (or recovers) from anthrax; when it is not so warm, it succumbs to anthrax.

85. *Problem:* What is the relation between industry and democracy, on the one hand, and between militarism and undemocratic government on the other?

Hypothesis: Industrialism favours democracy, militarism is hostile to it.

Evidence: The whole argument is an illustration of the *Deductive-Inductive Method.* The first paragraph is inductive, the second is deductive.

Inductive Evidence: (1) Athens and neighbouring States were essentially similar, except that Athens was industrial and democratic, while the others were neither. *Method of Difference.* (2) The Hanse Towns, the Low Countries, Norway, the U.S.A., Great and Greater Britain show societies different in many ways, yet all industrial and all democratic. *Method of Agreement.* (3) In England, as a whole, democracy has historically grown more and more as industrialism increased. *Method of Concomitant Variations.* (4) As regards the different parts of England, the industrial regions promoted democracy, the others did not. *Method of Difference.*

Deductive Evidence: Here it is shown from the nature of the case, in the light of what is already known about human nature, that industrial activities tend to produce democratic habits and dispositions, whereas military pursuits tend in the opposite direction.

86. *Problem:* To account for the high civilization of the Mayas in spite of the present unfavourable climatic and physical conditions of their country; and conversely, for the lower state of civilization of their neighbouring kinsmen in a tract of land where the climatic and physical conditions are now more favourable.

First Hypothesis (1): The suppositions brought together under (1) are all highly improbable, because all contrary to actual experience. They are not *veræ causæ.*

Second Hypothesis (2): This also is contrary to experience, and speculative.

Third Hypothesis (3): The suggested change of climate is a *vera causa,* as shown by the case of Palestine. If true, it would make the case of the Mayas and their neighbouring kinsmen very like normal cases.

87. *Problem:* What is the effect of a tropical climate on white people?

Hypothesis: It produces degeneration.

Evidence: The whole illustrates the *Deductive-Inductive Method.* The first half is inductive; the second half (beginning "And naturally so") is deductive.

Inductive Evidence: The same sort of people (racially) went to Canada and to the Bahamas. In the Bahamas they degenerated; in Canada they did not. The only relevant difference is the tropical climate of the Bahamas. *Method of Difference* (group form).

Deductive Evidence: Here it is shown from the nature of the case, in the light of what is already known of human nature, how a relaxing climate leads to physical and mental lassitude, and so to degeneration. The prophecy is highly speculative.

[Compare this with Nos. 62 and 82 in which the view opposed is that climatic and physical conditions are the *whole* cause of national or racial differences. In the present passage the question turns on the effect of a hot climate on a race accustomed to a more stimulating one.]

88. *Hypothesis:* The fear of war is a necessary spur to human progress.

Evidence for: Not stated. Probably the coincidence in certain countries of both war-preparations and works of progress. If so, a feeble use of the *Method of Agreement.*

Evidence against: If the hypothesis were true then countries very busy with war-preparations should show progress; but some of them don't. *Method of Agreement with negative result.* Even more obviously, countries not in fear of war should not be very progressive; yet some of them are. *Method of Difference* with negative result.

The inductive evidence against the hypothesis is next followed by *deductive reasoning* from the nature of the case in support of the same conclusion—namely, that fear of war leads to the diversion of a nation's resources from works of progress to those of destruction, and so actually retards progress.

The whole illustrates the *Deductive-Inductive Method*. Since the inductive part here precedes the deductive argument, it is an instance of Mill's Historical Method.

89. *Problem:* The relation between good government and freedom to criticize the Government.

First Hypothesis: Free criticism is fatal to good government.

Evidence for: Deductive argument from the nature of the case, namely, free criticism lowers the Government in the esteem of the people, and so makes it powerless.

Evidence against: If hypothesis were true, then the absence of criticism should make for good government. But Turkey is an obvious exception which seems fatal to the hypothesis. Moreover, the evidence for is inconclusive. Criticism may mend, not end, a Government (analogy of a doctor bleeding in order to improve a patient); and reducing the power of a Government is not the same as making it powerless—a powerless Government may be bad, but a Government may be all the more efficient for not being all-powerful.

Second Hypothesis: Free criticism is a condition of good government.

Inductive Evidence: Compare Turkey with

Britain. Britain with its Parliament and Press is efficient; Turkey without both is inefficient. *Method of Difference.*

Deductive Evidence: Human nature being what it is, criticism is necessary to keep the Government up to the mark. Nor is it unreasonable to expose the Government to free criticism. Other servants of the State, soldiers for instance, have to submit to far greater dangers in the State's interest.

An illustration of the *Deductive-Inductive Method.*

90. *Problem:* To account for the periodic changes of brightness of the star Algol.

Hypothesis: Algol is partly eclipsed periodically by a moon revolving round it.

Objection: Why is not the outline of the shadow of Algol's moon seen on Algol during such partial eclipses, as happens, e.g., during the partial eclipse of the sun by our moon?

Answer to Objection: Algol is so far away that it can only be seen as a point of light, not as a disc. Consequently a partial eclipse can only betray itself by a diminution in brightness.

Later Verification: In the light of certain laws of Mechanics and Physics the deductions described in the passage were drawn from the hypothesis, and inductively verified. The whole argument can be regarded as an example of the *Deductive-Inductive Method.*

By analogy the hypothesis is extended to other similar cases.

91. *Problem:* To account for the acid and the alkali on the poles of the electric machine during the decomposition of water. The problem is a *residual problem*—oxygen and nitrogen were expected, but nothing else.

First Hypothesis: They may come from the glass.

Verification: Eliminate the glass (by substituting gold vessels) and then, if the hypothesis is true, there should be no acid or alkali; but it made no difference. *Method of Difference with negative result.*

Second Hypothesis: The acid and alkali may come from foreign matter in the water.

Verification: Eliminate all foreign matter (by using distilled water), and then the acid and the alkali should disappear; but they were still there to some extent. Again *Method of Difference with negative result* in the main, but not entirely.

Third Hypothesis: They are due to the perspiration from the hands touching the instruments, and decomposing in the way described. Note the deductive element in this hypothesis.

Verification: Eliminate the perspiration, and nearly all the acid and alkali disappear. *Method of Difference.* But there still is a small residuum of both.

Fourth Hypothesis: The residuum is due to impurities from the air.

Verification: Eliminate the air (by putting the

machine in an exhausted receiver) and there is no acid or alkali. *Method of Difference.*

The solution of the whole problem really involves the co-operation of the last three hypotheses, some of the original acid and alkali being due to each of the three suggested sources.

92. *Problem :* To account for the difference between the calculated and the observed velocity of sound. Obviously a *residual problem* suggesting the *Method of Residues.*

Hypothesis : The excess or residual velocity is due to the heat developed by the condensation of the vibrations conveying sound.

Verification : In the light of physical laws already established, the effect of the alleged heat could be calculated exactly. *Deduction.* And the deduction just fitted. Here we have a deductive confirmation of an inductive hypothesis based on the Method of Residues. Really a form of the *Deductive-Inductive Method.*

Incidentally it is also a further confirmation of the law of the development of heat by compression. Such a *consilience of inductions* is greatly valued in science.

GROUP L

PROBABILITY AND CHANCE

5. (*a*) $\frac{93}{100}$. (*b*) $1 - (\frac{7}{100})^3$. (*c*) $(\frac{93}{100})^2$.

6. (*a*) $\frac{1}{2}$. (*b*) $\frac{1}{16}$. (*c*) $1 - \frac{9}{16} = \frac{7}{16}$.

7. (*a*) $\frac{1}{55}$. (*b*) $\frac{18}{55}$.

8. (i) $\frac{4}{5}$. (ii) $\frac{3}{5}$. (iii) $\frac{1}{5}$.

9. (i) $\frac{1}{4}$. (ii) $\frac{1}{16}$. (iii) $\frac{1}{16}$.

10. (i) $\frac{2}{5}$. (ii) $\frac{2}{5} \times \frac{1}{4} = \frac{1}{10}$. (iii) $\frac{3}{5} \times \frac{2}{4} \times \frac{1}{3} = \frac{1}{10}$.
[Explain on *general* grounds why (ii) = (iii) here.]

12. (i) $\frac{2}{3} \times \frac{2}{3} = \frac{4}{9}$. (ii) $(\frac{2}{3})^3 = \frac{8}{27}$. (iii) $1 - \frac{8}{27} = \frac{19}{27}$.

13. (i) $\frac{93}{100} \times \frac{94}{100}$. (ii) $1 - (\frac{93}{100} \times \frac{94}{100})$.
(iii) $1 - (\frac{93}{100} \times \frac{94}{100} \times \frac{95}{100})$.

15. (*a*) There are 15 ways in which one prize, and 20 in which the other may be awarded. Therefore 300 ways altogether, as they are independent. The probability of any one combination is consequently $\frac{1}{300}$.

(*b*) (i) If both prizes may not be awarded to the same person, then there are two prizes to be divided among 20, and each contributor has a chance of $\frac{2}{20} = \frac{1}{10}$. [Alternatively, probability of getting 1st prize $= \frac{1}{20}$. Probability of getting 2nd prize depends on chance of missing 1st prize

($=\frac{19}{20}$) and then $=\frac{1}{19}$, so that probability of getting 2nd prize $=\frac{19}{20}\times\frac{1}{19}=\frac{1}{20}$. Therefore total probability $=\frac{1}{20}+\frac{1}{20}=\frac{1}{10}$.]

(ii) If the same person may get both prizes, then the chance of getting 1st prize (*with* or without the 2nd) $=\frac{1}{20}$, and that of getting the 2nd (*with* or without the 1st) $=\frac{1}{20}$. Total $\frac{1}{10}$. But this counts *twice over* the chance of getting *both* prizes. This double chance has a separate value of $\frac{1}{400}$ ($\frac{1}{20}\times\frac{1}{20}$). As this has been counted twice in the gross total, the real total probability is $\frac{1}{10}-\frac{1}{400}=\frac{39}{400}$.

16. The watch will show the right time only once in twelve hours, i.e. in 720 mins. The probability is $\therefore \frac{1}{720}$.

17. (*a*) $\frac{1}{7}$. (*b*) $\frac{1}{49}$.

18. $\frac{1}{10}$.

19. (i) $\frac{7}{10}\times\frac{1}{2}=\frac{7}{20}$. (ii) $1-(\frac{3}{10}\times\frac{1}{2})=\frac{17}{20}$.
(iii) $1-(\frac{3}{10}\times\frac{1}{2}\times\frac{7}{10})=\frac{179}{200}$.

20. (i) $\frac{2}{10}\times\frac{4}{10}=\frac{2}{25}$.
(ii) $\frac{3}{10}$. [Average for all classes.]
(iii) $1-(\frac{8}{10}\times\frac{6}{10})=\frac{13}{25}$.
(iv) $1-(\frac{8}{10}\times\frac{6}{10}\times\frac{7}{10})=\frac{42}{125}$.

21. (i) $(\frac{2}{5})^3$. [More correctly, $\frac{2}{5}\times\frac{399}{999}\times\frac{398}{998}$.]
(ii) $1-(\frac{4}{5})^2=\frac{9}{25}$. [Or $1-(\frac{4}{5}\times\frac{799}{999}$.]
(iii) $1-(\frac{4}{5})^3=\frac{61}{125}$. [Or $1-(\frac{4}{5}\times\frac{799}{999}\times\frac{798}{998}$.]

22. The percentage of right conjectures is certainly very high—65 as against an expected 2. But a series of 100 attempts is too small to

warrant any inference. Moreover, there might have been collusion of various kinds.

23. By taking the mean we allow errors of too much and of too little to cancel each other. When accepting a particular observation we accept it with its error, whatever it may be.

24. *Misinterpretation of statistical laws* or regularities. They are merely averages taken over certain periods; they are summaries of the past, not calendars for the future. Their continuation is dependent on certain conditions, even if the relevant conditions are not always known completely. These conditions may change (social and other conditions do change), and the statistical averages will then change with them. There is no *must* about them; there is only a probability that, in the absence of marked relevant changes, they will hold good approximately for a while.

25. This legal dispute is a study in cross purposes due to the ambiguity of the word "accident." Of the many meanings of the term the following concern us here: (*a*) a calamity; (*b*) something not designed; (*c*) something not brought about by any cause, but just by "chance." Apparently the county court judge in Ireland only took into account meaning (*a*), his main problem being whether the calamity was the outcome of the victim's employment. His decision appears sensible; and the Irish Court of Appeal took the same view. Mr. Sankey took his stand on meaning (*b*), and

Lord Atkinson appears to have done so likewise. Lord Loreburn was attracted by meaning (*c*)—no wonder his metaphysics did not help him. Mr. Ronan appears to have returned to the common-sense view of the county court judge in Ireland; making, however, some concession to Mr. Sankey's use of the term accident (*b*), but refusing to treat the case as an " accident " in that sense

GROUP M

MISCELLANEOUS FALLACIES

1. *Fallacy of begging the question. Is* a *vacuum* nothing? That is just the question. If there is such a thing as a vacuum, two bodies will not touch, even if there is nothing more between them than an empty space.
2. *Fallacy of division.*
3. It is a gross misapplication of the term "normal" to anything human, without taking into account the proportion of men concerned.
4. *Fallacy of equivocation. Veritas,* or "truth," sometimes means "what accords with the facts," and sometimes "what is not a lie" (or what is "truthful"). A man in his cups may well be truthful, without being correct in his assertions.
5. *Fallacy of composition.*
6. *Fallacy of division.* To be one of "the best eleven" (= best team as a team) it is not necessary to be one of "the eleven best" individual players. What is true of the team need not be true of its individual members.
7. The sentence "Section B can be taken in

September only" is misinterpreted as though it meant "only Section B can be taken in September," whereas the first sentence states clearly that Section A can also be taken in September.

8. Ambiguous use of the phrase "I cannot see anything." It may mean "I am blind," and it may mean "There is nothing here for me to see" (or "I can see that there is nothing here").
9. *Fallacy of composition.* Every man must die; but all men need not die together.
10. *Fallacy of equivocation* (and of *four terms*). The first "not to be healthy" means "something different from being healthy"; the second means "to be ill."
11. *Fallacy of equivocation.* The "necessity" with which a man of perfect character does what is right is identical with free choice, not different from it; "necessity," when opposed to free choice, means external compulsion.
12. *Fallacy of confusing the contrary of a term with its contradictory*—"full of good" with "empty of evil" (= non-evil).
13. *Fallacy of equivocation.* "Some things," in the first line, means "various things," not the same kind of things. "What happens every day," in the fifth line, means "things of the same kind which happen every day." Also *four terms.*
14. *Fallacy of equivocation.* Money is said to be "plentiful" when people are ready to lend

it. Money may be plentiful or scarce, although the amount of coin or bullion remains the same.

15. *Fallacy of ambiguous construction.*
The second sentence is only acceptable if "we know not to be true" means "we do not know to be true," but not if (as happens here) it is also taken to mean "we know to be untrue."

16. *Fallacy of division.*

17. First sentence *partly inaccurate*—the position of the first boy could not be improved that way. Second sentence, *Fallacy of composition.*

18. *Fallacy of equivocation.* "Some" shares, in the first line, means "certain" shares (viz. those that will be drawn); in the second line it means "one or more" shares.

19. *Fallacy of composition.*

20. *Feeble analogy.* Analogy alone is never conclusive; here it is particularly feeble.

21. Self-contradiction. After his last shilling there will be no remainder to protect.

22. Warning against the fallacy of *consequens* (or affirming the consequent).

23. Warning against the *fallacy of affirming the consequent* (or illicit conversion).

24. Warning against the *fallacy of illicit conversion.*

25. Warning against the fallacy of *consequens* (or affirming the consequent).

26. *Fallacy of illicit conversion* (or of affirming the consequent).

27. Warning against *fallacious inference by added*

determinants. Though a doctor (or an actor) is a man, a bad doctor (or actor) need not be a bad man. *Bad* has a different meaning in the two cases.

28. *Fallacy of undistributed middle term.*
29. *Fallacious inference by added determinants.* To be *new as a member of the Council* is not the same as to be *new as a member of the staff*—an old member of the staff may be newly added to the Council.
30. Warning against the *fallacy of illicit conversion.*
31. *Fallacious inference by complex conception.* There need be no absolute diminution of the wealth of the country—only a diminution of the gold and silver, which may be more than compensated for by what is obtained in exchange.
32. *Fallacious inference by added determinants.* New dukes need not be new members; old members may have been promoted.
33. Second sentence is *self-contradictory*—the journey is to take "24 hours" and yet "no time at all." In the case supposed there may appear no change in clock-time as one passes from place to place, but time passes all the same, as may be seen by not interfering with one's watch from start to finish; and on returning to the starting-point it would be seen that a day had gone.
34. *Fallacious inference by added determinants.* Compare No. 27.
35. *Fallacy of composition.*

36. *Fallacy of undistributed middle term.*

37. *Illicit distribution of the major term.*

38. The argument assumes that "suppressing part of the truth" and "shutting one's eyes to unpleasant facts" mean the same thing. But they do not, and the syllogism commits the fallacy of four terms.

39. SaP ∴ $\bar{S}eP$ (or $\bar{S}a\bar{P}$). *Illicit inversion.*

40. *Fallacy of undistributed middle term.*

All rational opinions are shared by many PaM ;
Some prejudices are shared by many . SiM ;
∴ Some prejudices are rational . . SiP.

41. Confusion between "need not" and "should not"—what need not be done, may yet be the right thing to do.

42. *Fallacy of undistributed middle term.*

All just wars are popular . . . . PaM ;
The late war was popular . . . . SaM ;
∴ The late war was just . . . . SaP.

43. If you pay a doctor's bill immediately, he will think he has underestimated your value . . . . If A, then C ;
[But you don't want him to think so not C ;]
∴ You must not pay his bill immediately . . . . not A.

Valid ; but the major premise may not be true.

44. *Fallacy of affirming the consequent.* Even if "neither poor nor stupid" could be interpreted as the equivalent of "very clever or very rich," which it cannot be.

45. *Fallacy of composition.* False assumption that if one opponent is dangerous, two are equally,

or even more, dangerous. Two opponents may split the opposition vote, and so make it easier.

46. Because it is equivalent to the argument :
"If this hypothesis is true it can be verified ;
This hypothesis is verified ;
Therefore, this hypothesis is true."

47. *Fallacy of affirming the consequent.*

48. *Fallacy of undistributed middle term.*

49. *Fallacy of illicit distribution of the major term.*

50. *Fallacy of illicit conversion* of "all responsible beings are intelligent" into "all intelligent beings are responsible." The degree of intelligence of certain men and of all horses may not be such as to make them responsible.

51. *Fallacy of composition.* What is true of each event may not be true of the totality constituting the universe.

52. *Fallacious dilemma* through omission of other possible alternatives—the Government, e.g. may not mean to adopt any particular policy until advised by the Commission.

53. The error of the unscientific temper, which strives to shape the evidence to fit its opinions, instead of fitting its hypotheses to fit the facts.

54. Fallacy of *cum hoc ergo propter hoc.*

55. Fallacy of *cum hoc ergo propter hoc.* Ignoring the fact that there were fewer potential criminals in the country.

56. Fallacy of *post hoc ergo propter hoc.* The force

of the gas may have been spent by the time it came near the parapet.

57. Fallacy of *cum hoc ergo propter hoc.*
58. Fallacy of *post hoc ergo propter hoc.* The fact that it was a rising spur is obviously a relevant factor.
59. Fallacy of *post hoc ergo propter hoc,* based on a misapplication of the Method of Difference, very important circumstances being overlooked when comparing the earlier and the subsequent state of Germany.
60. Fallacy of *post hoc ergo propter hoc.* Other influences than protection may have been operative.
61. " Ought " shows a misinterpretation of a statistical frequency. It might have been expected as probable, but there is no " ought " about it. Moreover, the frequency only holds good " in the long run "—small groups (as in this case) may well show a difference.
62. Ambiguous use of " follows." A cause is an antecedent in *time.* Whether its effect is *spatially* in front or behind is a matter of indifference.
63. If even trained specialists are fallible, how much more so will untrained laymen be. The errors are evidence of the need, not of less, but of more training.
64. Warning against the fallacy *post hoc ergo propter hoc,* with an illustration by way of analogy.
65. Question begging.
66. Circular argument.

67. Question begging. The "reason" given is only a more positive way of stating what is first stated rather negatively.

68. *Fallacy of equivocation.* The sense in which a budget can be objectively described as "socialistic" does not imply all the evils here associated with it. To make it imply all this without proof is simply *begging the question.*

69. A vast circulation is no evidence of correctness—fallacy of *non causa pro causa* (irrelevant *reason,* not cause). The second sentence is an evasion of the real issue—*ignoratio elenchi.*

70. *Fallacy of irrelevant reason* (*non causa pro causa*).

71. *Argumentum ad baculum*—abusing your opponents.

72. What the experiment verified was that the destruction of the diseased lungs helped to prevent the spread of the disease among the healthy cattle. This might be due to all sorts of reasons more consonant with known facts than the fanciful disease-devils. In such cases more particularly, to regard verification as a *proof* of the hypothesis is a fallacy of *consequens.*

73. This is a warning against the fallacy of ignoring special circumstances (*a dicto simpliciter ad dictum secundum quid,* and the converse fallacy of *a dicto secundum quid ad dictum simpliciter*).

74. Fallacy *a dicto secundum quid ad dictum simpliciter.* What happens when education is

uncommon may cease to happen when education is so widespread as to cease to be a distinction.

75. Fallacy *a dicto simpliciter ad dictum secundum quid. Normally* freedom may be a natural right. But *circumstances alter cases*, and a man may forfeit his freedom by his action.

76. *Fallacy of division.* Even if we grant that the Early Church, taking all its customs as a whole, was a model society, it does not follow that each of its customs or institutions (such as the absence of private property) was perfect.

77. What is given as a reason is only a repetition of what was said before. What can be observed needs no reason, and should be expressed simply in a statement of fact, instead of in the specious form of an argument.

78. Rash generalization based on the fallacy of *cum hoc ergo propter hoc*—if a State is corrupt *and* a Republic, that alone does not justify the conclusion that it is corrupt *because* it is a Republic.

79. Fallacy of *a dicto simpliciter ad dictum secundum quid.* Civil servants enjoy special privileges, and may have special obligations.

80. *Fallacy of rash generalization.* Those that have been deterred by the fear of imprisonment would naturally not appear in court to proclaim the fact.

81. *Circular argument.*

82. Warning against the simple conversion of SaP into PaS.
83. A careless application of the Methods of Concomitant Variations and of Difference. The relation established is that between light and the *appearance* of colours, while the conclusion refers to the colours themselves. From the evidence cited here one is not justified in supposing that "colours" and their "appearances" are the same thing.
84. Immediate inference by complex conception (or added determinants), falsified by ambiguity—"Time is money" being taken too literally.
85. A specious application of the Joint Method of Agreement and Difference. Certainly the cause of the prosperity of the prosperous countries will be something found there and not in other countries; but the simple converse of this is not true. A country may be prosperous *in spite* of certain institutions rather than *because* of them—hence the common fallacy of *cum hoc ergo propter hoc*.
86. Fallacy of *cum hoc ergo propter hoc*; and a dubious teleological argument—what gratifies one's instincts is not always right.
87. Here we see excessive respect for authority prompting extravagant speculation.
88. Here we have a jocular warning against far-fetched conclusions from the flimsiest data. There are plenty of learned books containing theories which are quite as extravagant if not quite so amusing.

89. At best the analogy makes it somewhat probable that the average length of human life might be longer than it is. But no evidence is adduced in support of the suggestion that our meat-eating habits are at fault—not even tables of the comparative ages to which carnivorous and herbivorous animals live!

90. An interesting retort, but not conclusive. It is plausible so long as merely vague terms like "sometimes" are employed. The proper test would be one based on exact statistical enumeration, which would probably show no correlation between the sorcerer's methods and results, and at least some degree of correlation between the methods and results of trained doctors.

GEORGE ALLEN & UNWIN LTD
London: 40 Museum Street, W.C.1
Cape Town: 73 St. George's Street
Sydney, N.S.W.: Wynward Square
Wellington, N.Z.: 4 Willis Street